CONCRETE ADMIXTURES

CONCRETE ADMIXTURES

Vance H. Dodson, Ph.D.

STRUCTURAL
Engineering
series

VNR VAN NOSTRAND REINHOLD
_____ New York

Library of Congress Catalog Card Number 90-41627
ISBN 0-442-00149-5

Printed in the United States of America

Van Nostrand Reinhold
115 Fifth Avenue
New York, New York 10003

Chapman & Hall
2-6 Boundary Row
London SE1 8HN, England

Thomas Nelson Australia
102 Dodds Street
South Melbourne, Victoria 3205, Australia

Nelson Canada
1120 Birchmount Road
Scarborough, Ontario M1K 5G4, Canada

16   15   14   13   12   11   10   9   8   7   6   5   4   3   2

Library of Congress Cataloging-in-Publication Data

Dodson, Vance H.
   Concrete admixtures / Vance H. Dodson.
      p.   cm.—(VNR structural engineering series)
   Includes bibliographical references and index.
   ISBN 0-442-00149-5
   1. Cement—Additives.   I. Title.   II. Series.
TP884.A3D63   1990
666'.94—dc20   90-41627
                                                            CIP

To
My wife, three children, and three grandsons

—— VNR STRUCTURAL ENGINEERING SERIES ——

Structural Engineering Theory

STRUCTURES: FUNDAMENTAL THEORY AND BEHAVIOR by Richard Gutkowski

STRUCTURAL DYNAMICS, 2nd Ed., by Mario Paz

MICROCOMPUTER-AIDED ENGINEERING: STRUCTURAL DYNAMICS by Mario Paz

EARTHQUAKE RESISTANT BUILDING DESIGN AND CONSTRUCTION, 2nd Ed., by Norman Green

THE SEISMIC DESIGN HANDBOOK edited by Farzad Nacim

Steel Design

STEEL DESIGN FOR ENGINEERS AND ARCHITECTS by Rene Amon, Bruce Knobloch and Atanu Mazumder

Concrete Design

HANDBOOK OF CONCRETE ENGINEERING by Mark Fintel

STRUCTURAL DESIGN GUIDE TO THE ACI BUILDING CODE, 3rd Ed., by Paul F. Rice, Edward S. Hoffman, David P. Gustafson and Albert J. Gouwens

TORSION OF REINFORCED CONCRETE by T. Hsu

MODERN PRESTRESSED CONCRETE, 3rd Ed., by James R. Libby

Masonry

SIMPLIFIED MASONRY SKILLS by R. Kreh

Wood

MECHANICS OF WOOD AND WOOD COMPOSITES by Jozsef Bodig and Benjamin Jayne

STRUCTURAL USE OF WOOD IN ADVERSE ENVIRONMENTS edited by Robert W. Meyer and Robert M. Kellogg

STRUCTURAL DESIGN IN WOOD by Judith J. Stalnaker and Ernest C. Harris

Tall Buildings

DEVELOPMENTS IN TALL BUILDINGS by the Council on Tall Buildings
ADVANCES IN TALL BUILDINGS by the Council on Tall Buildings
SECOND CENTURY OF THE SKYSCRAPER by the Council on Tall Buildings

Other Structures-Related Books

ICE INTERACTION WITH OFFSHORE STRUCTURES by A. B. Cammaert and D. B. Muggeridge
FOUNDATION ENGINEERING HANDBOOK by Hans Winterkorn and H. Y. Fang
FOUNDATION ENGINEERING FOR DIFFICULT SUBSOIL CONDITIONS by Leonardo Zeevaert
ADVANCED DAM ENGINEERING edited by Robert Jansen
TUNNEL ENGINEERING HANDBOOK by John Bickel and T. R. Kuesel
SURVEYING HANDBOOK edited by Russell Brinker and Roy Minnick

CONTENTS

PREFACE AND ACKNOWLEDGMENTS

This book is written for an audience which includes, among others (1) concrete producers who use (or do not use) admixtures to describe what can be expected by the use of the fifth ingredient in their day-to-day batching of concrete, (2) concrete contractors who look for certain placing and finishing characteristics of concrete, in spite of the environmental temperature and humidity, and (3) engineers and architects who are concerned about the properties of the finished and hardened concrete. The underlying theme throughout the book is one that says that admixtures cannot make bad concrete good concrete, but can, when properly used, make good concrete better concrete.

This treatise often cites my own research and field experiences of 28 years in the cement and concrete industries and my observations and conclusions may conflict with those of others in these fields of endeavor. In addition, I have ignored a vast number of articles published in the literature, especially those whose results are either too complicated or clouded with controversy. I basically like to generalize and try to look at the practical side of things and make complicated issues as simple as possible so that the man on the street can understand them. Having raised three children and participated in the upbringing of three grandchildren, my two philosophies of life are (1) avoid complications and (2) don't fix it if it ain't broke.

I would like to acknowledge Mr. J. David Dunham who struggled with my poor penmanship in putting the hand-written word into a typed, readable form, the Construction Products Division of W. R. Grace for giving me access to its extensive library facilities and the many technicians, chemists and group leaders that I often drove "to the edge" in collecting data in experiments that seemed to be almost as crazy as their supervisor.

Chapter 1

PORTLAND CEMENT

INTRODUCTION

In preparing this chapter, the author has chosen what he considers the most logical, reasonable, and simple approaches to the manufacture and chemistry of portland cement. The author quotes a man he considers to be much wiser than himself, "The phenomenon by which a suspension of particles of portland cement in water evolves into the solid cement paste binder of mortar or concrete, and by which this binder may be modified by interaction with environmental factors, is still poorly understood" [1]. Twenty years later, considerable controversy in explaining how portland cement "does its thing" when mixed with water still exists. The author's goal in this chapter is to express, in rather simple terms, how portland cement reacts with water and how the resulting mass hardens in the presence and/or absence of admixtures.

First, it should be agreed on what the author is talking about when the subject of portland cement is addressed. It is, by definition, a hydraulic cement, produced by pulverizing clinker consisting essentially of hydraulic calcium silicates, usually containing one or more forms of calcium sulfate as an interground addition [2].

The author has intentionally underlined several terms in the definition of a portland cement to clarify its nature. The term *hydraulic cement* refers to one that hardens by a chemical interaction with water and is capable of doing so under water [3]. Portland cement *clinker* is a partially fused mass primarily composed of *hydraulic calcium silicates* [4]. The *calcium silicates* are essentially a mixture of tricalcium silicate ($3CaO \cdot SiO_2$) and dicalcium silicate ($2CaO \cdot SiO_2$) in which the former is the dominant form.

Calcium sulfate occurs in two forms in nature, gypsum ($CaSO_4 \cdot 2H_2O$) and natural anhydrite ($CaSO_4$) and is usually added as the former or as a mixture of the two, in which the former is the predominant species. An *interground addition* is a material that is either ground or blended in limited amounts into hydraulic cement clinker, such as calcium sulfate, just described, or air entraining agents, grinding aids, set retarders, etc., during the manufacture of portland cement.

Three other terms require defining:

1. Portland cement paste: a mixture of portland cement and water (sometimes referred to as neat cement paste).
2. Portland cement mortar: a mixture of portland cement, fine aggregate (usually sand), and water.
3. Portland cement concrete: a mixture of portland cement, fine aggregate, coarse aggregate, and water.

It has taken the author years to convince his children, grand children, and spouse that what they called "cement" highways, sidewalks, and driveways should really be referred to as "concrete."

HISTORY

The debate over who first discovered portland cement still rages. The Greeks, Romans, and others built great and durable structures but from materials not related to portland cement. The most probable source of the early cement materials was calcined limestone (lime) and a pozzolanic material, such as volcanic ash, which contained an active form of silicon dioxide (SiO_2). In the presence of moisture, chemical reactions took place between the three components which caused cementitious processes to occur and were not the result of the calcium silicate hydraulicity observed with portland cement.

The development of portland cement is generally credited to Joseph Aspdin, an English brick layer, who in 1824 was awarded a patent for his product [5]. Aspdin named his product portland cement, because after mixing it with water and sand, the hardened product had a color resembling that of the limestone quarried on the Isle of Portland in the English Channel. As early as 1760, an English engineer, John Smeaton, produced a material similar to that of Aspdin but discarded the lumps (clinker) formed in the hottest region of his kiln because they were difficult to pulverize [6]. (*This is analogous to throwing out the baby with the bath water.*)

At best, Aspdin's product was far inferior in performance to the portland cement manufactured in the years that followed. His scanty records show that he crushed limestone and heated it to convert it to lime. He then blended it with clay by grinding the two to a fine slurry in water. After drying, the slurry solids were further heated in a furnace similar to a lime kiln. During the course of heating, the raw materials combined to form a clinker which contained hydraulic calcium silicates and aluminates. Finally, Aspdin ground the product from his kiln to create the finished product.

The crudeness of Aspdin's product can be attributed to a number of factors: (1) lack of knowledge of the proper mix proportions of the lime and clay, (2) relatively low kiln temperatures, and (3) insufficient grinding of the final product to a fineness that would permit the hydraulic compounds to react with water and develop strength at a reasonable rate. When the industry attributes the invention of portland cement to Aspdin, it is not correct in the true sense of the word. He really only invented portland cement clinker. It was not until several years later that it was discovered that calcium sulfate had to be interground with the portland cement clinker to control the rate of hydraulicity and time of setting.

The first portland cement, as is known today, to be made in the U.S. was produced by David Saylor at Coplay, PA, in 1871, in vertical kilns, similar to those used in the calcination of limestone. The increasing demand for both quantity and quality of portland cement led to the introduction of the rotary kiln in the U.S. in 1889. This type of kiln had been developed by Frederick Ransome, in England in 1885. Over the years that followed, numerous changes and refinements in the manufacture of portland cement have occurred, and today it is the most often used construction material and is the lowest cost building material, on a pound for pound basis.

MANUFACTURE

Raw Materials

The four principal elemental oxides needed to make portland cement clinker are calcium oxide (CaO), silicon dioxide (SiO_2), aluminum oxide (Al_2O_3) and iron oxide (Fe_2O_3). The most common natural occurring sources of these major oxides are listed in Table 1-1. Calcium oxide is not found as such in nature but is formed during the cement clinkering process when calcium carbonate is heated. This conversion is often referred to as calcination and is basically what Aspdin did in the first step of his process.

Clinker Process

After crushing to 3/4″ size, or smaller, each of the raw materials that are available to a given portland cement producer is transferred to separate storage facilities. Each is then chemically analyzed, and the calculated amounts of each are ground to a fine powder, combined and thoroughly blended to produce the required type of portland cement clinker. The grinding and blending operations are often combined and are called "raw grinding." The blend is then either fed directly into the clinker kiln or into a preheater (or series of preheaters) prior to its entry into the kiln. Most cement producers today employ the preheater concept because it improves the productivity and energy efficiency of the kiln operation. The preheater utilizes the hot gases leaving the feed end of the kiln, and as a result the moisture

Table 1-1.   Raw Materials Used in the Manufacture of Portland Cement Clinker.

OXIDE	NATURAL SOURCE	COMPOSITION
Calcium Oxide (CaO)	Calcite	90–95% Calcium Carbonate
	Limestone	80–90% Calcium Carbonate
	Aragonite	90–95% Calcium Carbonate
	Sea Shells	80–90% Calcium Carbonate
	Marl	35–65% Calcium Carbonate
Iron Oxide (Fe_2O_3)	Limonite	80–86% Iron Oxide
	Clay	6–8% Iron Oxide
	Shale	9–10% Iron Oxide
Silicon Dioxide (SiO_2)	Sand	90–95% Silicon Dioxide
	Shale	50–60% Silicon Dioxide
	Clay	45–50% Silicon Dioxide
	Marl	16–20% Silicon Dioxide
Aluminum Oxide (Al_2O_3)	Bauxite	30–75% Aluminum Oxide
	Clay	38–42% Aluminum Oxide
	Shale	20–25% Aluminum Oxide

and most of the carbon dioxide, from the mineral carbonates, are driven off before the raw materials enter the kiln. Prior to the preheater innovation, the energy laden gases were allowed to escape into the environment. What the author has just simply and briefly described is the so-called "dry" process for producing portland cement clinker.

In the "wet" process, the blending and grinding operations are carried out with the quantified amounts of raw material in the form of a water slurry. In most other respects, the "dry" and "wet" processes are the same. The promoters of the "wet" process claim that a more homogeneous blend of raw materials is achieved by the slurry technique and thus a more uniform finished clinker is produced. The advocates of the "dry" process maintain that more energy is required to drive off the moisture in the slurried product.

The author next considers what happens when the raw mix (whether it be produced by the "wet" or "dry" process) makes its way through the kiln. The cylindrically shaped kiln is tilted downward approximately 3 to 5 degrees from entrance to exit and is rotated along its semi-horizontal axis (hence the name, rotary kiln) causing the feed to pass from one end to the other at a speed determined by its length, downward slope, and rate of rotation. The temperature at the hotter end of the kiln is in the range of 2700° F to 2800° F and is created by the injection of burning gas, oil, or pulverized coal. Most kilns today are fired by the last because of the cost and availability of the other two.

As the raw material feed passes from the relatively cool end towards the burning hell at the other end, changes occur in its components. In the absence of preheaters, the free and combined moisture in the raw mix components are driven off in the temperature range of 200° F to 300° F (which is accomplished in a preheater, if it is a part of the kiln system). As the mix passes through a temperature zone; that is in the range of 1300° F to 1800° F, calcination of the carbonates take place (again, most of this will occur in a preheater). The so-called burning zone is in the range of 1900° F to 2800° F. It is in this temperature zone that chemical reactions take place that lead to the formation of the hydraulic components of the clinker. In the hottest region (2600° F to 2800° F), approximately 20% to 30% of the mass is in the liquid state, and it is in this mixture of liquid and solids that the principal chemical reactions occur. The aluminum and iron oxides are necessary in the manufacture of the clinker because they act as fluxing agents to the mix. If some liquid were not formed during the burning process, the reactions leading to the formation of the hydraulic calcium silicates would be much slower, would require higher temperatures, and probably not be complete. During their fluxing action, the oxides of aluminum and iron also form hydraulic compounds by combining with calcium oxide, which play important roles in determining the properties of the finished product.

The rotary action of the kiln causes the semi-fluid mass to form small balls, ranging from $1/2''$ to $2''$ in diameter, which are dumped from the hot end of the kiln and air cooled. The hot air from the cooler is either injected back into the hot end of the kiln or passed on to the preheater(s). The original raw materials no longer exist as such but have been converted to portland cement clinker which is dark grey in color and, after cooling, is made up of complex compounds. The four principal elemental oxides, mentioned earlier, that have no hydraulic properties, now exist in chemical combinations which do possess those characteristics.

Those clinker components having hydraulic properties are very simply (the author emphasizes the word simply) described in Table 1-2. None of the clinker phases is pure. For example C_4AF is really a solid solution of components ranging from C_6A_2F to C_6AF_2 with C_4AF being its mean composition. Also, alkali ions and magnesium ions are often incorporated in one or more of the four phases. The relatively pure phases can, however, be made under carefully controlled conditions in the laboratory and are often referred to as pure or synthesized phases or compounds.

Not all of the raw materials entering the kiln comes out as clinker. For every 100 pounds of raw material fed to the kiln, only approxi-

Table 1-2. Portland Cement Clinker Compounds.

CHEMICAL NAME	MINERAL PHASE NAME	CHEMICAL FORMULA	CEMENT CHEMISTS' DESIGNATION[a]
Tricalcium Silicate	Alite	$3CaO \cdot SiO_2$	C_3S
Diacalcium Silicate	Belite	$2CaO \cdot SiO_2$	C_2S
Tricalcium Aluminate	Celite	$3CaO \cdot Al_2O_3$	C_3A
Tetracalcium Alumino-ferrite	Iron	$4CaO \cdot Al_2O_3 \cdot Fe_2O_3$	C_4AF

[a]The cement chemist prefers to express the chemical components in an abbreviated form; for example, $C=CaO$, $A=Al_2O_3$, $S=SiO_2$, and $F=Fe_2O_3$.

mately 85 pounds exit as clinker. The remainder is swept out of the kiln as dust, moisture, and carbon dioxide. The air containing these ingredients is passed through filters, and the solids collected on the filter is referred to as kiln dust. In many cases the dust is returned to the kiln either with the raw feed or by insufflation (which involves blowing the dust into the hot end of the kiln). The dust is usually high in alkali (mainly as sulfates), and its return to the kiln results in a build up of alkali in the clinker. In other cases, the dust is either blended with clinker during its grinding or with the finished cement prior to its storage. Whatever the case may be, the finished product is going to have a higher alkali content than that before the portland cement producer was forced by environmental agencies to collect his kiln dust. The effect of the increasing alkali content on the properties of concrete will be discussed in the chapters that follow. Coincidentally, shortly after the pollution regulations were enforced upon the cement industry, the problem of "acid rain" surfaced. Since the kiln dust is highly alkaline, it could have, when allowed to escape into the atmosphere, served to neutralize some of the acids generated by other industries (mainly acid oxides of sulfur and nitrogen).

Probably the most important phase of the portland cement clinker is the tricalcium silicate (C_3S). The dicalcium silicate (C_2S) phase is formed rather rapidly during the burning process, but its conversion to C_3S is slow and best obtained in the temperature range of 2700° F to 2800° F, and its rate of formation is affected by the chemical nature of the starting ingredients and the rate of heating of the raw mix [7].

Grinding Process

Now, considering the portland cement clinker—those little grey colored balls that John Smeaton discarded because they were so difficult to pulverize. After cooling, the clinker is ground, or pulverized, in what is called a finish mill along with a relatively small amount of gypsum (or a mixture of gypsum and natural anhydrite). The finish mill is, like the kiln, cylindrical in shape and rotated about its central axis but not tilted downward. The material entering the finish mill is aided in passing from one end to the other by lifters or fins, welded to the interior walls of the mill which raise the mass, including the grinding media, to a certain height as the mill rotates from which the mixture falls to the bottom of the mill, and the clinker is crushed. As a result, the particle size of the gypsum and clinker is progressively reduced as they pass through the finish mill. The grinding media are spherical in shape and made from a special steel alloy.

Although the finish mills will vary from plant to plant, they are usually multi-compartmental, with different size grinding media in each compartment. The compartments are separated by steel bulkheads which have designed sized perforations which allows the pulverized material to pass through to the next compartment but restrain the passage of the grinding media in that particular compartment.

At the discharge end of the finish mill, the ground material is conveyed to an air separator where, after separation, (1) the fines are sent to the finished product storage silo, (2) the coarse particles are returned to the grinding mill for further particle size reduction, and (3) the very fine portion is swept by air to a series of dust collectors. A typical finish grinding mill system is very simply diagrammed in Figure 1-1. The fineness of the finished cement is such that 85% or more of it will pass the No. 325 (45 μm) sieve which has 105,625 openings per in.^{2}, and almost all of it will pass through a No. 200 (75 μm) sieve, which has 40,000 openings per square inch. The fine-

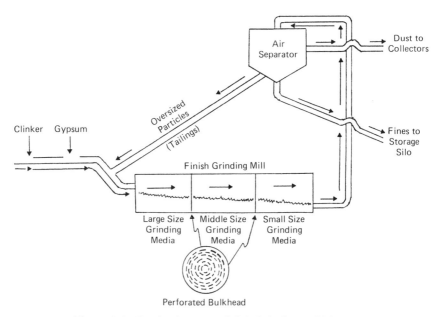

Figure 1-1. Portland cement finish Grinding Mill System.

ness of portland cement is most often expressed in units of cm^2/g as measured by either air permeability or turbidimetry [8][9].

Finish grinding is a science in itself. In closed circuit grinding operations, such as that diagrammed in Figure 1-1, the air separator and finish mill cannot be considered as separate unrelated entities. They operate as a team whose individual characteristics must be combined in such a way as to attain maximum operating efficiency of the total circuit. Achieving this proper balance between the two has been described in the literature [10][11][12].

TYPES

Not all portland cements are alike. They differ in their composition and fineness, depending upon their intended use. The *five* principal portland cements are described in ASTM C150 [13]. These are designated in the U.S. as types I, II, III, IV, and V. Comparable standards have been established in Canada, but the cements are called types 10, 20, 30, 40, and 50, respectively. The five types of cement have certain behavioral characteristics which lend themselves to a variety of uses (Table 1-3). The classification shown in Table 1-3 is based essentially on the amounts of C_3A and C_3S in the cement and its fineness. The first three types were, at one time, widely produced in which an air entraining addition was added during its finish grinding. These were (and still are) labeled types I-A, II-A, and III-A portland cements. Since most air entrained concrete in today's construction market is produced by the addition of an air entraining admixture (see Chapter 6) during the batching of the concrete, very little type

Table 1-3. Behaviorial Characteristics of Portland Cement.

TYPE	CHARACTERISTICS[a]
I	Most widely produced and used for general construction.
II	Possesses moderately low heat and a moderate degree of resistance to sulfate attack.
III	Sets and gains strength rapidly, often referred to as "high early."
IV	Has a low heat of hydration and is often used in massive structures where thermal cracking might occur.
V	Has a high sulfate resistance and is useful in those areas where concrete is exposed to an environment rich in metal sulfate salts.

[a]The characteristics of the types II through V are based on (or compared with) those of the type I cement.

"A" cement is specified or produced. The average chemical composition of the five types of portland cement, expressed in terms of their chemical phases is summarized in Table 1-4.

All four phases of portland cement evolve heat when they react with water (and calcium sulfate), and this chemical reaction, which leads to hardening and development of strength is called *hydration*, and the heat generated during the reaction is called *heat of hydration*. So that one can better appreciate the influence of the various phases of portland cement on its properties (Tables 1-3 and 1-4), Table 1-5 lists the heat of hydration of each phase, synthetically prepared in the laboratory.

It should be pointed out that the heats of hydration of the C_3A and C_4AF phases were measured in the absence of calcium sulfate and, therefore, are somewhat higher than that experienced in portland cement. The method used to measure the heat of hydration of portland cement (or its synthetically prepared phases) is described in ASTM C186 [14]. A comparison of the heats of hydration of the first four types of portland cement at several time periods following their being mixed with water are shown in Table 1-6.

Table 1-4. Average Phase Composition of Portland Cement.

CEMENT TYPE	AVERAGE COMPOSITION—%				
	C_3A	C_4AF	C_3S	C_2S	TOTAL[a]
I	12	9	55	20	96
II	7	12	45	30	94
III	12	8	65	10	95
IV	5	13	25	50	93
V	3	14	40	35	92

[a]The percentages do not total 100 because other compounds; i.e., calcium sulfate, magnesium oxide, alkali sulfates, etc. are not included.

Table 1-5. Heat of Hydration of Synthesized Portland Cement Phases.

CHEMICAL PHASE	HEAT OF HYDRATION—Btu/lb
C_3A	516
C_4AF	172
C_3S	211
C_2S	98

Table 1-6. Average Heat of Hydration of Portland Cement.

AGE—DAYS	TYPE I	TYPE II	TYPE III	TYPE IV
	HEAT OF HYDRATION OF CEMENT—Cal/gram[a]			
3	62.5	47.5	75.0	40.0
7	78.5	62.5	90.0	50.0
28	95.0	80.0	102.5	65.0
90	102.5	87.5	107.5	75.0

[a]In order to convert cal/gm to Btu/lb multiply by 1.80.

The reason for type II being classed as a moderate heat concrete (with respect to a type I) is its lower C_3A and C_3S contents, and these two phases are responsible for most of the heat generated during the cement's hydration. Type IV portland cement has an even lower heat of hydration because of its low C_3A and C_3S content. Because its C_3A content is considerably lower than that of a type I, the type V produces less reaction products that are susceptible to sulfate attack by its environment. Type III undergoes rapid set and develops higher early strength than a type I, because of its higher fineness.

FINENESS

Fineness, which is an expression of the amount of surface area of the cement that is immediately available for reaction with water, is very important to the rate of hydration of cement as well as its rate of strength development in paste, mortar, and concrete. An example of the latter phenomenon is illustrated in Table 1-7. In his work, the author prepared mortars in accordance with ASTM C109, from a type I cement ground to various increasing fineness values in a laboratory

Table 1-7. Effect of Fineness on Compressive Strength of Mortar.

CEMENT FINENESS cm²/g[a]	1-DAY	3-DAY	7-DAY	28-DAY
	COMPRESSIVE STRENGTH—PSI			
3550	1900	3650	4425	5630
4195	2430	4200	4695	5767
4840	3350	4350	5030	5850
5485	4225	4820	5310	6050

[a]Reference [8]

steel ball mill [15]. While the particle size (or fineness) of portland cement is important to its rate of hydration as well as its rate of strength development, the particle size distribution within the cement is just as critical. The author found that portland cements of equal fineness develop higher 1 day and 28 day compressive strengths as the particle size distribution narrows. This finding is in agreement with results published in the literature [16]. His experience has shown that a wide particle size distribution leads to erratic (1) early strengths and gains in strengths with time, (2) time of setting, and (3) water demand to produce the necessary flow of paste and mortars and slump of concrete. The particle size distribution, assuming that the particles are spherical in shape, for three types of portland cement is shown in Table 1-8. The values represent *the average* for a large number of cements of each type by the sedigraphic method and are expressed in microns (μm) to the nearest whole number.

The author would like to point out here that the air permeability test for fineness was developed by R. C. Blaine, an Associate Materials Research Engineer at the National Bureau of Standards and often referred to as the Blaine fineness [8]. The turbidimetric method was developed by L. A. Wagner and often referred to as the Wagner fineness [9]. The fineness as determined by the Wagner method is always of a lower magnitude than that found by the Blaine method, but the two are somewhat related. For example, 25 cements manufactured in the 1970s, and chosen at random, exhibited an average Blaine:Wagner fineness ratio of 1.80, with a standard deviation of 0.10 and a coefficient of variation of 5.8%. Twenty-five cements manufactured over the period 1985 to 1987, also chosen at random, had an average Blaine to Wagner ratio of 1.85, with a standard de-

Table 1-8. Typical Particle Size Distribution of Portland Cement.

PARTICLE SIZE DIAMETER μm[a]	PARTICLE SIZE—% GREATER THAN TYPE OF PORTLAND CEMENT		
	I	II	III
45	91	94	97
7	27	29	39
1	3	3	5
0.6	1	1	2

[a]1 μm = 4×10^{-5} in.

viation of 0.14 and a coefficient of variation of 7.6%. Perhaps advances in finish grinding and in air separation technologies over that space of years are responsible for the differences in the average ratios.

Probably the most frequently performed test for the fineness of portland cement, because of its simplicity, is the measurement of the amount of the cement that passes the No. 325 (45 μm) sieve [17]. However, the results of the test must be considered in conjunction with either the Blaine or Wagner fineness values for it to be meaningful. When the percentage passing the sieve is plotted as a function of the Blaine fineness, a rather scattered pattern is obtained as evidenced in Figure 1-2. The best straight line drawn through such data points (by computer) has a slope of 26.5. Over the range of Blaine fineness of 3500 to 3750 cm^2/g (a difference of 250 cm^2/g) the percent passing the sieve varies from 84.8% to 95.0%, a difference of 10.2 percentage points. When the range of Blaine values is 3750 to 4000 cm^2/g (again, a difference of 250 units), the amount passing the sieve only differs by 4.3 percentage points. The conclusion is that the finer the cement is ground, the closer the correlation will be between the two fineness values. The differences in the rates that portland cements undergo hydration, harden, and develop strength not only depends upon their fineness but on other factors, such as (1) chemical compositions, (2) ambient temperatures, and (3) the amounts of water in the cement-water systems (or concentration of the cement in the cement paste).

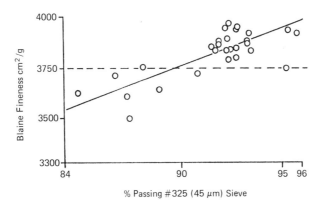

Figure 1-2. Relationship between Blaine Fineness and percent passing #325 (45 μm) sieve.

HYDRATION PROCESSES

A large amount of literature has been generated which deals with the early hydration reactions of portland cement that lead to the development of a structure which causes stiffening, hardening and finally, strength. Numerous mechanisms have been proposed on the basis of results observed on pure, synthesized cement phases, but trying to apply those mechanisms to portland cement has been found to be difficult to do. When portland cement is mixed with water, the resulting paste begins to stiffen after a period of time and undergoes initial set, final set, and development of compressive, flexural and tensile strength. All of these processes are the result of the build-up of hydration products which tend to consolidate the mass and hold it together. There are two accepted methods for measuring the time of setting of portland cement paste, and only one is recommended for that of concrete [18][19][20]. All three methods are based on the resistance of the mass under test to the penetration of a weighted probe.

Aluminate Phases

The reaction of the tricalcium aluminate (C_3A) phase of portland cement is a two step process.

Step 1. Immediate reaction with water and calcium sulfate forming a compound called ettringite—$3CaO \cdot Al_2O_3 \cdot 3CaSO_4 \cdot 32H_2O$.

Step 2. The ettringite coating the underlying tricalcium aluminate then reacts with it and additional water to form a "low sulfate" ettringite—$3CaO \cdot Al_2O_3 \cdot CaSO_4 \cdot 12H_2O$.

In the absence of calcium sulfate, or sulfate ions, the tricalcium aluminate reacts directly with water to form $3CaO \cdot Al_2O_3 \cdot 6H_2O$. This is a very rapid and exothermic reaction that leads to a quick hardening known as "flash set."

The reaction of the tetracalcium alumino ferrite (C_4AF) phase is not as clear cut as that of tricalcium aluminate because of the complexity of the phase. Compounds analogous to ettringite and its low sulfate form are thought to be produced in which some of the aluminum ions are partially replaced with small amounts of ferric ions

(Fe^{+3}) along with some amorphous ferrite phases [21]. In the absence of calcium sulfate (or sulfate ions), tri- and hexahydrates are thought to form, but their rate of formation is slower than that of the direct hydration of tricalcium aluminate.

Silicate Phases

Most of the extensively reported studies on the hydration of portland cement have been focused on the hydration of the tricalcium silicate (C$_3$S) phase, because it is the major component. There is considerable evidence that it undergoes hydration in five steps [22]. These can be best illustrated by the nature of its isothermal conduction calorimetric curve shown in Figure 1-3. The C$_3$S was synthesized in the laboratory and a water-C$_3$S ratio of 0.52 was used in preparing the paste for the calorimeter study.

Step 1. When tricalcium silicate is mixed with water there is an immediate reaction between the two. Evidence for this immediate reaction is the initial exotherm in the isothermal conduction calorimetric curve. This is further indicated by the rapid supersaturation of the aqueous phase with respect to calcium hydroxide and the decided rise in its pH. It is widely accepted that during Step 1 the surface of the tricalcium silicate reacts with water to form calcium hydroxide and a hydrated calcium silicate less rich in calcium oxide. The exotherm produced by pure tricalcium silicate is much smaller than that created by a portland cement (see Figure 1-6).

Step 2. An induction, or dormant period in the hydration of the tricalcium silicate, which covers a time span of about four (4) hours occurs. The product of the immediate hydration in Step 1 coats the

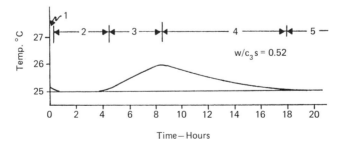

Figure 1-3. Isothermal conduction calorimetric curve of a synthetic C$_3$S and water mixture.

underlying unhydrated silicate and because it has a low water permeability it retards further reaction with water. At the end of Step 2 the depth of the hydrated coating on the original grains of the tricalcium silicate has been estimated to be in the range of 20 to 100 Å ($1\text{Å} = 3.9 \times 10^{-9}$ inches).

Step 3. The rate of tricalcium silicate hydration suddenly accelerates and proceeds through a maximum. At the beginning of Step 3, solid calcium hydroxide begins to crystallize from its supersaturated condition in the aqueous phase to produce a saturated solution. This seems to trigger an acceleration in the hydration and a renewed acceleration in the exothermic reaction. This phenomenon is commonly attributed to a rather abrupt change in the nature of the hydration product which exposes fresh surfaces of unhydrated tricalcium silicate particles. Dormant periods followed by acceleration periods in other systems are quite common [23]. As shown in Figure 1-3, the dormant period ended after about 4-1/2 hours and the rate of heat evolution peaked after approximately 8-1/2 hours. The author has found that the pH of the aqueous phase of a tricalcium silicate-water paste drops at a time approximating the end of the dormant period. The results of another study have shown that solid calcium hydroxide begins to form after approximately four hours of hydration and thereafter crystallization proceeds rapidly [24]. The matter of what occurs at the end of Step 2 and the beginning of Step 3 is analogous to the question of what came first, the chicken or the egg. Some claim that the sudden crystallization of the calcium hydroxide is due to the abrupt increase in tricalcium silicate hydration rate. Others hold to the theory that the crystallization causes the acceleration of the tricalcium silicate hydration.

Step 4. The rate of hydration of the tricalcium silicate slowly decreases over a period of 15 to 20 hours and no further surge in reaction rate is observed. At the end of Step 3, sufficient hydration products have been formed that further reaction becomes controlled by the rate of diffusion of water through them to the anhydrous inner core material and hydration slows down, during which time the temperature of the paste asymptotically approaches that of the calorimeter. During Step 3 about 20% of the tricalcium silicate in portland cement undergoes hydration and approximately the same amount hydrates during Step 4 but over a longer period of time (Figure 1-4) [25].

Step 5. The tricalcium silicate continues to react slowly accompanied by a slow heat liberation.

The chemical reaction between tricalcium silicate and water in its most simple form, is pictured in Equation 1-1.

$$2[3CaO \cdot SiO_2] + 6H_2O \rightarrow 3CaO \cdot 2SiO_2 \cdot 3H_2O + 3Ca(OH)_2 \quad (1\text{-}1)$$

This equation is only approximate because the calcium silicate hydrate product is poorly crystalline and is a non-stoichiometric solid material. The ratio of CaO to SiO_2 and the amount of water associated with the hydration product is still uncertain. The results of the analysis of tricalcium silicate pastes hydrated for 1 to 30 years have shown that the hydration produces a compound having a mean calcium oxide-silicon dioxide ratio of 1.46, with a range of 1.2 to 1.8 [26]. More recently it has been reported that the preferable calcium oxide- silicon dioxide-ratio in the tricalcium silicate hydration is 1.4 to 1.5, at least in the early ages of hydration [27]. Regardless of its chemical composition, the hydrate is very closely related to the naturally occurring mineral tobermorite which has the formula $5CaO \cdot 6SiO_2 \cdot 5H_2O$ and takes its name from Tobermory, Scotland, where it was first discovered. Hence the product of the hydration is frequently referred to as tobermorite.

Although dicalcium silicate, C_2S, can exist in four well established forms, a, a', β and γ, the form in which it normally occurs is the β variety. Rapid cooling of the clinker or the incorporation of small

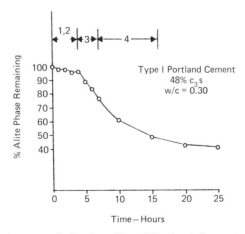

Figure 1-4. Reaction rate of alite in a Type I Portland Cement—Water Mixture.

amounts of certain metallic oxides into its structure prevents it from converting to the y form which has no hydraulic properties [28].

β-C_2S, the second most abundant phase in portland cement clinker hydrates much like C_3S but is more sluggish in its reaction than C_3S because its component ions are more densely packed in its structure. The structure of C_3S has "holes" in it, is more water permeable, and less thermodynamically stable. The most probable formula for the hydration product of β-C_2S is $5CaO \cdot 6SiO_2 \cdot 5H_2O$ or tobermorite [29].

After reviewing the experimental data, of which a great part is conflicting, in the literature, one has to wonder if the formulae of the hydration products of C_3S and C_2S are really that important. This is why the author often refers to the reaction products, whatever their true formulae might be, as the "glue" that holds the paste, mortar, and concrete together.

Knowing the formulae of the hydrated silicates is important in one respect in that it allows one to determine the potential amounts of calcium hydroxide that can be produced by the two in concrete. If the product of hydration of C_3S is $3CaO \cdot 2SiO_2 \cdot 3H_2O$, as shown in Equation 1-1 and that of β-C_2S is tobermorite (Equation 1-2), the amounts of hydrated silicates and liberated calcium hydroxide can be calculated, using simple stoichiometry.

$$6[2CaO \cdot SiO_2] + 12H_2O \rightarrow 5CaO \cdot 6SiO_2 \cdot 5H_2O + 7Ca(OH)_2 \quad (1\text{-}2)$$

The results of that mathematical exercise are listed in Table 1-9. The quantities of calcium hydroxide produced by the hydrating silicates are important in a number of ways. First, with time, it can be leached, by alternate wetting and drying, from the cement paste, mortar or concrete and become the principal source of efflorescence, a white, unsightly surface stain. Second, that portion of the calcium hydroxide that is not leached out will eventually react with the carbon dioxide

Table 1-9. Products of the Complete Hydration of C_3S and β-C_2S.

CALCIUM SILICATE 100 lbs	$C_3S_2H_3$ (lb)	$C_5S_6H_5$ (lb)	$Ca(OH)_2$ (lb)
C_3S	75	—	47
C_2S	—	71	50

in the permeating atmosphere and can be converted to calcium carbonate. This chemical conversion has a negative effect on the volume of the hardened mass and leads to what is often called "carbonation" shrinkage. Third, calcium hydroxide, contributes no strength to the cementitious mass, and in the case of concrete can detract from its strength because it tends to deposit between the paste and the aggregate and weaken the bond between the two. This phenomenon is illustrated in Figure 1-5, wherein the plate-like crystals of the calcium hydroxide are clearly shown in the microphotograph adjacent to an aggregate socket in the concrete specimen. Fourth, the presence of the liberated calcium hydroxide is an important factor in the pozzolanic reaction that occurs when certain mineral admixtures are added (Chapter 7) and when air entraining admixtures are used (Chapter 6).

Before closing this chapter, the author would like to point out that the isothermal conduction calorimeter curve for a portland cement is quite similar to that of pure, synthesized C_3S (Figure 1-3) because the C_3S phase is by far the predominant one present. A typical curve for a type I cement is shown in Figure 1-6. When that same portland

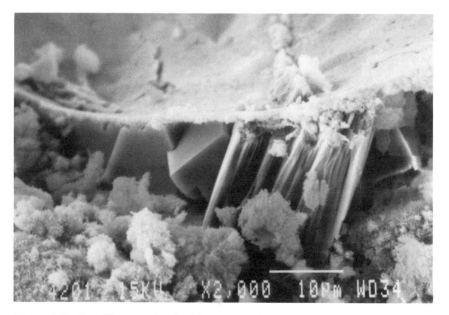

Figure 1-5. Plate-like crystals of calcium hydroxide deposited between concrete matrix and aggregate socket.

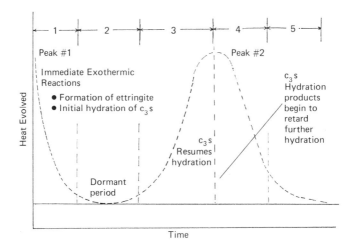

Figure 1-6. Isothermal conduction calorimeter curve of a Type I Portland Cement.

cement does not contain sufficient calcium sulfate (often referred to as being "sulfate starved"), a third peak in the temperature vs. time curve occurs, as shown in Figure 1-7. The consequences of having too little calcium sulfate in a given cement will be discussed in subsequent chapters.

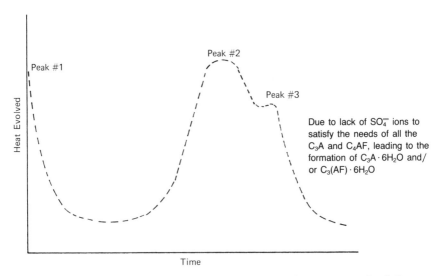

Figure 1-7. Isothermal conduction calorimetric curve of a Type I Portland Cement containing insufficient calcium sulfate.

In case one is wondering why the author began this text with a chapter devoted to portland cement when there are a number of learned texts on the subject available, he explains that the brief and simple summary of the science was presented in order to lay the ground work for the chapters that follow. After all, it is the portland cement and/ or its hydration products that determine the efficiency with which an admixture performs.

REFERENCES

[1] Mather, B., "Portland Cement—Research, Testing and Performance," *Journal of Materials*, Vol. 5, No. 4, pp. 832–841 (1970).

[2] ASTM C150, "Standard Specification for Portland Cement," *Annual Book of ASTM Standards*, Vol. 04.01, pg. 134 (1986).

[3] "Cement and Concrete Terminology," *American Concrete Institute, Publication SP-19*, pg. 72 (1988).

[4] "Cement and Concrete Terminology," *American Concrete Institute, Publication SP-19*, pg. 106 (1988).

[5] Aspdin, J., "Artificial Stone," *British Patent No. 5022* (1824).

[6] Bogue, R. H., "Portland Cement," *Cooperative Investigations by the Portland Cement Association and the National Bureau of Standards*, Paper No. 53, pp. 411-431 (1949).

[7] Odler, I., Dorr, H., "Tricalcium Silicate Formation by Solid State Reactions," *Ceramic Bulletin*, Vol. 56, No. 12, pp. 1086–1088 (1977).

[8] ASTM C204, "Standard Test Method for Fineness of Portland Cement by Air Permeability Apparatus," *Annual Book of ASTM Standards*, Vol. 04.01, pp. 208–215 (1986).

[9] ASTM C115, "Standard Test Method for Fineness of Portland Cement by the Turbidimeter," *Annual Book of ASTM Standards*, Vol. 04.01, pp. 139–148 (1986).

[10] Mardulier, F. J., "Balance—Key to Mill/Separator Operation," *Rock Products*, June (1969).

[11] Mardulier, F. J., Wightman, D. L., "Efficient Determination of Mill Retention Time, Parts I, II and III," *Rock Products*, June, July, August (1971).

[12] Dodson, V. H., Welch, P. W., Mardulier, F. J., "Finish Grinding—The Grace Umbrella Technique, Part I and II," *Rock Products*, June, July, pp. 72–75, 92–96 (1974).

[13] ASTM C150, "Standard Specification for Portland Cement," *Annual Book of ASTM Standards*, Vol. 04.01, pp. 153–159 (1986).

[14] ASTM C186, "Standard Test Method for Heat of Hydration of Hydraulic Cement," *Annual Book of ASTM Standards*, Vol. 04.01, pp. 185–191 (1986).

[15] ASTM C109, "Standard Test Method for Compressive Strength of Hydraulic Cement Mortars (Using 2 in. or 50 mm.) Cube Speciments," *Annual Book of ASTM Standards*, Vol. 04.01, pp. 74–79 (1986).

[16] Locher, F. W., Sprung, S., Korf, P., "The Effect of Particle Size Distribution on the Strength of Portland Cement," *ZEMENT-KALK-GIPS*, No. 8, pp. 349–355 (1973).

[17] ASTM C430, "Standard Test Method for Fineness of Hydraulic Cement by the 45 μm (No. 325) Sieve," *Annual Book of ASTM Standards*, Vol. 04.01, pp. 279–281 (1986).

[18] ASTM C191, "Standard Test Method for the Time of Setting of Hydraulic Cement by Vicat Needles," *Annual Book of ASTM Standards*, Vol. 04.01, pp. 204–207 (1986).

[19] ASTM C266, "Standard Test Method for Time of Setting of Hydraulic Cement Paste by Gillmore Needles," *Annual Book of ASTM Standards*, Vol. 04.01, pp. 242–245 (1986).

[20] ASTM C403, "Standard Test Method for Time of Setting of Concrete Mixtures by Penetration Resistance," *Annual Book of ASTM Standards*, Vol. 04.02, pp. 210–213 (1988).

[21] Fukuhara, M., Goto, S., Asaga, K., Daimon, M., Kondo R.," Mechanisms and Kinetics of C_4AF Hydration with Gypsum," *Cement and Concrete Research*, Vol. II, No. 3, pp. 407–414 (1981).

[22] Kondo, R., Ueda, S., *Proceedings of the Fifth International Symposium on the Chemistry of Cement: Part II*, October 7–11, pp. 203–255 (1968).

[23] Kondo, R., Daimon, M., "Early Hydration of Tricalcium Silicate: A Solid Reaction with Induction and Acceleration Periods," *Journal of the American Ceramic Society*, Vol. 52, No. 9, pp. 503–508 (1969).

[24] Greene, K. T., "Early Hydration Reactions of Portland Cement," *Fourth International Symposium on Chemistry of Cement*, Washington, DC, pp. 359–374 (1960).

[25] Angstadt, R. L., Hurley, F. R., Private Communication to V. H. Dodson, W. R. Grace Research Center, (1969).

[26] Lachowski, E. E., Mohan, K., Taylor, H. F. W., Lawrence, C. D., Moore, A. E., "Analytical Electron Microscopy of Cement Pastes Hydrated for Long Times," *Journal of the American Ceramic Society*, Vol. 64, No. 6, pp. 319–321 (1981).

[27] Li, S., Roy, D. M., "Preparation and Characterization of High and Low CaO/SiO_2 Ratio—Pure C—S—H— for Chemically Bonded Ceramics," *Journal of Material Research*, pp. 380–386 (1988).

[28] Lea, F. M., "The Chemistry of Cement and Concrete," Chemical Publishing Company, Inc., New York, NY 3rd Edition, pg. 45 (1970).

[29] Shebl, F. A., Helmy, F. M., "Tobermorite as the Final Product of β-C_2S Hydration," *Cement and Concrete Research*, Vol. 15, pp. 573–580 (1985).

Chapter 2

ADMIXTURES—GENERAL CONCEPTS

CLASSIFICATION

An admixture is defined as a material other than water, aggregate, and hydraulic cement which might be added to concrete before or during its mixing. This is not to be confused with the term, addition, which is either interground with or blended into a portland cement during its manufacture. An addition, is classified as being either (1) a processing addition, which aids in the manufacture and handling of the finished product, or (2) a functional addition which modifies the use properties of the cement.

Admixtures can function by several mechanisms.

1. Dispersion of the cement in the aqueous phase of concrete.
2. Alteration of the normal rate of hydration of the cement, in particular the tricalcium silicate phase.
3. Reaction with the by-products of the hydrating cement, such as alkalies and calcium hydroxide.
4. No reaction with either the cement or its by-products.

Those that function via the first two mechanisms are called *chemical admixtures* in order to differentiate them from the others that perform by the last two mechanisms. The performance of chemical admixtures is specified in ASTM C494, and only some of their effects on the physical properties of concrete are summarized in Table 2-1 [1]. Other requirements include length change (shrinkage), flexural strength, and relative durability factor. These will be discussed in subsequent chapters.

The author first considers compressive strength which is defined as the measured maximum resistance of a concrete (or mortar) specimen

Table 2-1.   Certain Physical Requirements of Concrete Treated with a Chemical Admixture.

REQUIREMENT	TYPE OF CHEMICAL ADMIXTURE[a]						
	A[b]	B[c]	C[d]	D[e]	E[f]	F[g]	G[h]
Water content, % of control-max	95	—	—	95	95	88	88
Time of setting, allowable deviation from control:hr:min							
Initial: at least	—	1:00 later	1:00 earlier	1:00 later	1:00 earlier	—	1:00 later
not more than	1:00 earlier nor 1:30 later	3:30 later	3:30 earlier	3:30 later	3:30 earlier	1:00 earlier nor 1:30 later	3:00 later

Final: at least not more than	— / 1:00 earlier nor 1:30 later	— / 3:30 later	1:00 earlier / —	—	1:00 earlier / —	— / 1:00 earlier nor 1:30 later	— / 3:30 later
Compressive strength, % of control-min							
1 day	—	—	—	—	—	140	125
3 days	110	90	125	110	125	125	125
7 days	110	90	100	110	110	115	115
28 days	110	90	100	110	110	100	110
1 year	100	90	90	100	100	100	100

[a] ASTM C494 [1].
[b] Water Reducing.
[c] Retarding.
[d] Accelerating.
[e] Water reducing and retarding.
[f] Water reducing and accelerating.
[g] Water reducing, high range.
[h] Water reducing, high range and retarding.

to axial compressive loading, expressed as force per unit cross-sectional area usually in terms of pounds per square inch (psi) or megapascals (MPa, 1 Mpa = 145 psi). The compressive strength of concrete is determined on cylindrical specimens whose diameter is one-half that of its length; i.e., $3'' \times 6''$, $4'' \times 8''$, or $6'' \times 12''$ or on portions of beams previously broken in flexure [2][3][4]. This later method is not intended as an alternative test for concrete cylinders, and the test values obtained by the two test methods are not interchangeable and not necessarily comparable. The compressive strength of mortar is determined on $2'' \times 2'' \times 2''$ cubes in accordance with ASTM C109 [5].

COMPRESSIVE STRENGTH

The compressive strength of concrete and/or mortar, in which admixtures or additions are absent, is a function of the following variables:

- Water-cement ratio, which is the ratio of the amount of water, exclusive of that absorbed by the aggregates, to the amount of cement, by weight, and is frequently expressed as W/C.
- Cement content in lb/yd^3, or cement factor, sometimes denoted as C.F.
- Temperature.
- Chemical composition and physical properties of the cement.
- Gradation of the aggregate.
- Ratio, by weight, of coarse to fine aggregate.
- Method of curing.

Abrams found that for a given set of materials and conditions, the compressive strength of concrete is dependent on the concentration of the cement in the cement paste [6]. This concept has come to be known as Abrams' Law.

Some of the data generated by the author 20 years ago, illustrating Abram's Law are shown in Figure 2-1. In this work a fixed amount of cement was used, 550 lb/yd^3 and the amount of coarse and fine aggregate was increased, at a fixed weight ratio of 55 to 45, as the water-cement ratio was reduced from 0.70 to 0.40 to maintain a constant yield of 27 ft^3/yd^3. Water-cement ratio vs. compressive

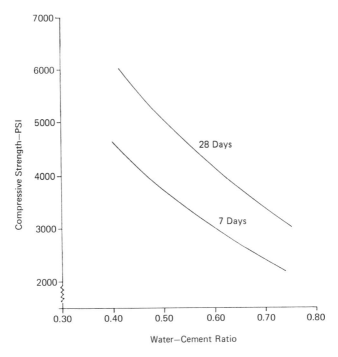

Figure 2-1. Effect of water-cement ratio on compressive strength of concrete, C.F. = 550 lbs.

strength curves having similar contours have been reported by many others [7].

Several years later the author examined concretes having a fixed water-cement ratio but with varying amounts of cement (400 to 600 lb/yd³). Plots of the compressive strengths at two ages are shown in Figure 2-2. As the cement content was increased—in 50 lb increments from concrete to concrete—the amounts of coarse and fine aggregates were reduced but held at a weight ratio of 55 to 45, respectively, to maintain a constant yield. It was found that this particular experiment was very difficult to complete and it was not only after four different attempts that the author settled for a water-cement ratio of 0.50 using the then currently available raw materials.

After that work was completed and the data were analyzed, the author was somewhat surprised to find that the contours of the curves in Figure 2-2 were very close to being mirror images of those in Figure 2-1. It was concluded that if all the aforementioned variables that

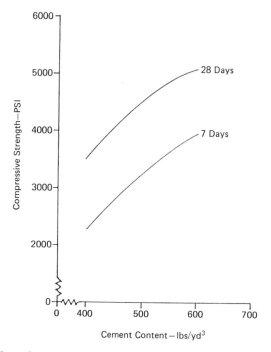

Figure 2-2. Effect of cement content on compressive strength of concrete at a constant water—cement ratio of 0.50.

play a part in determining compressive strength were held constant, except water-cement ratio (W/C) and cement content (C.F.), the compressive strength of concrete, at any age, should be directly relateable to a combination of these two variables. This relationship is expressed in Equation 2-1.

$$\text{Compressive Strength (at any age)} \simeq \frac{\text{C.F.}}{\text{W/C}} \qquad (2\text{-}1)$$

While he did not originally set out to prove this concept, Gruenwald generated data that supported this postulation much earlier, at least over the range of 300 to 500 C.F.s and 0.53 to 0.73 water-cement ratios [8]. The author has chosen to call the term C.F. ÷ W/C the Omega Index Factor, or O.I.F., of concrete and of mortar, in some instances.

Now, consider what happens when a series of concretes is fabricated in which both the water-cement ratio and cement content are varied (and the author has performed this exercise several hundred times). The results of only one series of laboratory concretes, chosen at random, are described in Table 2-2. As one proceeds from left to right in Table 2-2, it can be seen that as the C.F. was decreased, the amounts of both coarse and fine aggregates were increased to maintain the desired yield of 27 ft^3/yd^3, *but* the weight ratio of the coarse to fine aggregate was maintained at a weight ratio of 55 to 45. The slumps and yields of all three concretes were essentially the same. The compressive strengths of the concretes described in Table 2-2 are plotted as a function of the O.I.F.s in Figure 2-3. The linearity of the three plots indicates that there is a direct relationship between O.I.F. and compressive strength at the three ages of test. By dividing the compressive strengths of the three concretes by their C.F.s, an estimate of the strength contribution of the cement in each concrete can be calculated (Table 2-3).

After examining the strength contribution of a large number of different cements, over a wide range of water-cement ratios and cement factors, the author has concluded that those values can vary to a considerable extent, as shown in Table 2-4. It should be pointed out that the strength contributions cited in Tables 2-3 and 2-4 are for concretes that have been moist cured in the laboratory. It is of interest to note here that when numerous members of ASTM Committee C-9, responsible for concrete and concrete aggregates, were canvassed, the

Table 2-2. Concrete Designed to Test O.I.F. Concept.

CONCRETE NO.	1	2	3
Cement—lb/yd^3	552	479	403
Fine Aggregate—lb/yd^3	1410	1450	1475
Coarse Aggregate—lb/yd^3	1740	1790	1815
Water—lb/yd^3	294	296	295
Slump—in.	3-1/4	3-1/2	3-1/2
Air—%	1.6	1.4	1.5
W/C	0.533	0.618	0.732
O.I.F.	1036	775	550
Yield—ft^3/yd^3	27.01	27.15	27.07
Compressive Strength—PSI			
1-day	1325	1145	1005
7-days	3750	3110	2610
28-days	5790	4790	3870

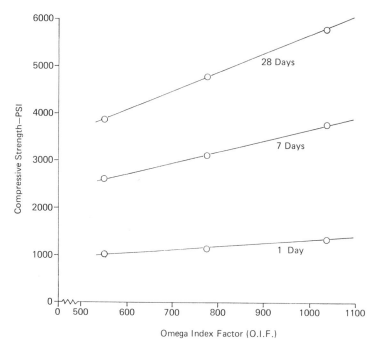

Figure 2-3. Relationship between compressive strength and Omega Index Factor of (O.I.F.) of Concrete.

author was not surprised that the average answer to his question as to what strength contribution to concrete could be expected from the average portland cement, in the absence of an admixture, was 10 psi / lb. cement /yd³ of concrete at 28 days.

The compressive strengths reported in the producer's cement mill certificate are measured on 2-in.³ mortar cubes whose composition is

Table 2-3.   Strength Contributed by Portland Cement to Concrete—PSI/lb of Cement/yd³ of Concrete.

| | STRENGTH CONTRIBUTION OF CEMENT AT VARIOUS AGES | | |
CONCRETE NO.	1-DAY	7-DAYS	28-DAYS
1	2.4	6.8	10.5
2	2.3	6.5	10.0
3	2.5	6.5	9.6
Average	2.4	6.6	10.0

Table 2-4. Strength Contribution of Portland Cements to Concrete—PSI/lb of Cement/yd³ of Concrete.

AGE—DAYS	STRENGTH CONTRIBUTION—RANGE
1	1–3
7	5–8
28	8–12

fixed; i.e., same weights of sand, cement, and water with a water-cement ratio of 0.48. Their method of fabrication, curing, and testing is described in ASTM C109 [5]. Their compressive strengths are only relateable to those reported in the previous mill certificates and are only a measure of the uniformity of the cement from one delivery to another and not relateable to either laboratory or field concretes made from the same cement because of the latter's wide difference in water-cement ratio, cement content, coarse to fine aggregate ratio and content, and temperature. Also, the concrete specimens prepared in the field are subject to a wide variation in temperature, humidity, and handling.

However, the O.I.F. concept also makes itself evident in laboratory prepared and cured mortar cubes. For example, three mortar mixes were fabricated in which various amounts, by *weight* of the cement in mortar No. 1 were replaced with equal volumes of the sand, used in ASTM C109, in Nos. 2 and 3 [5]. A description of the mortars is given in Table 2-5, and their compressive strengths, at the three ages of test, are plotted as a function of their O.I.F.s in Figure 2-4. When the mortars in the three mixes are converted to concrete having the same volume percent of mortar, the O.I.F. values, expressed in units of psi/gram of cement/liter of mortar, can be equated, by a series of mathematical manipulations, to O.I.F. values in units of psi/lb of cement/yd³ of concrete. There is fairly good correlation between the strength contributions measured in the two systems, although the variation in values exhibited by the three mortars, at any given age of test, is of some concern. This is attributed to the variation in the air contents of the three mortars, which were not measured.

The author has purposely devoted a great deal of discussion to the O.I.F. concept, because his intention is to show, in the several chapters that follow, how it can be used to evaluate the strength contributions of admixtures.

Table 2-5. Description of Mortars Used to Evaluate the O.I.F. Concept.

| | MORTAR MIX NUMBER | | |
INGREDIENTS AND PROPERTIES	1	2	3
Cement—grams[a]	740	592	444
Sand—grams	2053	2177	2302
Water—grams	359	359	359
Yield—liters	1.409	1.411	1.413
Water—cement ratio	0.48	0.61	0.81
Cement factor—g/L mortar	525	420	314
O.I.F.	1094	688	388
Compressive Strength—PSI			
3-day	3910	2260	995
7-days	5075	3025	1335
28-days	5600	3575	2150
Compressive Strength Contribution PSI/g cement/L mortar			
3-day	7.4	5.3	3.2
7-days	9.7	7.2	4.2
28-days	10.7	8.5	6.8
Compressive Strength Contribution PSI/lb cement/yd^3 concrete[b]			
3-day	6	5	3	Avg. 5 (6)[c]
7-days	8	6	4	6 (8)
28-days	9	8	6	8 (10)

[a]Type I cement.
[b]Values rounded off to nearest whole number.
[c]Values found for the same cement in concrete of the same mortar contents.

TIME OF SETTING

The time of setting of concrete, in the absence of admixtures, like compressive strength, should also be a function of its cement content and water-cement ratio. All things being equal; i.e., same cement, water, stone, sand, stone to sand weight ratio and temperature, the time of setting of concrete should be inversely related to the former and directly related to the latter. Again, the Omega Index Factor comes into play. Although the author has confirmed this reasoning many times in the laboratory, only one example will be illustrated here. A series of six concretes having different cement factors and various water-cement ratios were fabricated and their times of setting were measured in accordance with ASTM C403 [9].

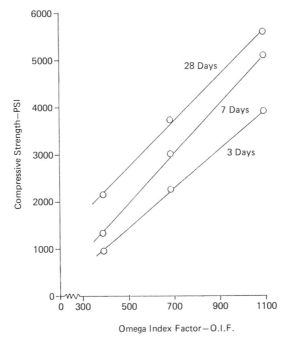

Figure 2-4. Omega Index Factor vs. Compressive Strength (PSI) of Mortars.

The pertinent concrete data are listed in Table 2-6, and plots of the initial and final times of setting of the concretes vs. their O.I.F.s are shown in Figure 2-5. The two sets of data points (Figure 2-5) approximate straight lines and the correlation factors for the initial and final time of setting time data are 0.982 and 0.987, respectively. It is of interest to note that a plot (Figure 2-6) of the differences in initial and final times of setting, listed in Table 2-6, vs. the initial times of setting is essentially linear, indicating that as the time of initial time of setting increases, the difference between the two increases. This phenomenon will be further illustrated in the chapters that follow.

The time of setting of portland cement is commonly determined on paste, using the Vicat apparatus [10]. In this method, a fixed amount of the cement and a sufficient amount of water to meet normal consistency requirements are prescribed [11]. The degree to which the paste develops stiffening is measured by the extent of penetration of the Vicat needle (1 mm in diameter) under a weight of 300 grams at various time intervals following the mixing period. A typical pen-

Table 2-6. Concretes Fabricated to Illustrate the Effect of O.I.F. on Time of Setting.

CONCRETE COMPONENTS AND PROPERTIES	CONCRETE NO.					
	1	2	3	4	5	6
Cement[a]—lb/yd^3	392	485	396	594	490	594
Water—lb/yd^3	333	333	273	324	268	268
Slump—in.	6-1/4	6-3/4	1-1/2	5-3/4	1-1/2	1
Air—%	1.0	0.8	2.1	1.2	2.0	2.0
W/C	0.85	0.69	0.69	0.55	0.55	0.45
C.F. ÷ W/C (O.I.F.)	461	703	574	1080	891	1320
Initial Set—min	345	325	325	295	305	265
Final set—min	460	400	425	340	390	300

[a]Type I cement.

etration vs. time curve for a Type I portland cement paste, in which the cement is *said* to exhibit normal set at standard laboratory conditions is shown in Figure 2-7. Initial and final set is proclaimed to have occurred when the penetration is 25 mm and 0 mm, respectively.

The procedure used to determine the time of setting of concrete is described in ASTM C403 [9]. In this method, the concrete under test is passed through a No. 4 sieve (1/4" openings) before being subjected to penetration by needles of various diameters, under laboratory conditions that are the same as those prescribed for testing the paste. When the resistance of the screened concrete to penetration reaches a value of 500 psi and 4000 psi, initial and final sets, re-

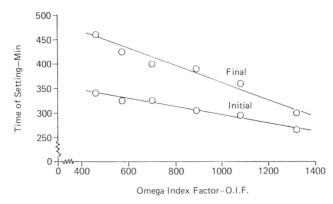

Figure 2-5. Relationship between time of setting of concrete and its O.I.F.

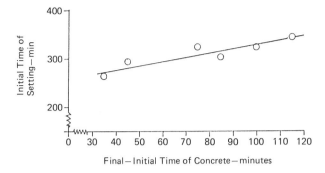

Figure 2-6. Influence of initial time of setting of concrete on its final time of setting.

spectively, are said to have occurred. The resistance to penetration is determined by dividing the force to produce a penetration to a depth of 1 in. by the surface area of the bearing surface of the needle. When the same Type I cement (Figure 2-7) is used in concrete (C.F. = 517 lbs, W/C = 0.53), the screened concrete's response to penetration is illustrated in Figure 2-8.

It was mentioned earlier in this discussion that the time of setting of the portland cement, described in Figures 2-7 and 2-8 was *normal*. Defining the term, normal, as it pertains to time of setting is somewhat ambiguous, but it is best described as the retention of plasticity by a paste, mortar, or concrete over a period of time required for proper and easy placing.

Several comments should be made here before leaving the subject.

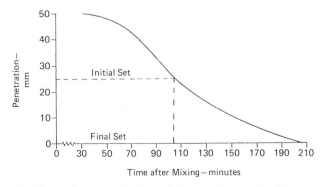

Figure 2-7. Time of setting of a Type I Cement Paste—by Vicat Apparatus.

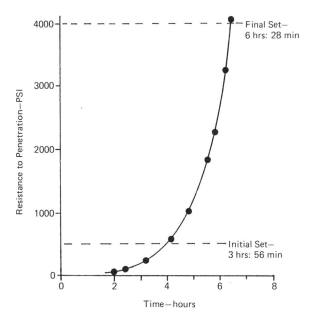

Figure 2-8. Initial and final time of setting of concrete.

1. The terms, initial and final (time of setting) are arbitrarily chosen limits of plasticity or resistance to penetration by a weighted probe.

2. In the case of portland cement concrete, the time of initial setting is an indication that the concrete mass will no longer respond to vibration and is ready for finishing. On the job site, this is noted by the disappearance of the glossy sheen on the surface of the concrete.

3. The time of setting of a given cement as determined by the paste method and that measured by the concrete method cannot be related because of the differences in cement factor and water-cement ratio, both of which have a profound influence, as illustrated through the use of the Omega Index Factor.

4. The time of setting of concrete is greatly influenced by temperature, and this will be elaborated on in several of the chapters that follow. Very little concrete is mixed and placed at standard laboratory conditions (73.4 ± 3° F, 50% R.H.). In addition, the mass of concrete plays an important role in determining its time of setting. In the laboratory, the volume of screened concrete approximates one gallon while one cubic yard of concrete in the field is equivalent to approximately 200 gallons. The larger the volume of the concrete, the shorter

its time of setting will be because the heat generated by the hydrating cement is slow to dissipate into its surroundings; and, as it accumulates in the body of the concrete, the time of setting is accelerated. (The author once followed a ready mix truck to a job site where a small 6″ deep slab was being placed, took a grab sample of the concrete as it came down the truck chute, screened it, placed the screened material in a gallon can and began the penetration tests, exposing the test specimen to the same environmental conditions as those to which the slab was exposed. The concrete finishers had done their job and moved on to another job site some 60 minutes before the sample exhibited initial set.)

Some will argue with the author, but he strongly feels that the initial time of setting of portland cement paste, mortar, or concrete is due in small part to the formation of ettringite (Chapter 1) but in most part to the result of the tricalcium silicate coming out of its dormant stage and forming relatively large amounts of its hydration product. The author is also of the opinion that the principal cause for final set is the additional amounts of C_3S hydration products that form following the initial set.

The importance of the O.I.F., as it pertains to time of setting, will be illustrated in subsequent chapters when the influence of various admixtures on this important physical property of concrete is discussed.

REFERENCES

[1] ASTM C494, "Standard Specification for Chemical Admixtures for Concrete," *Annual Book of ASTM Standards,* Vol. 04.02, pp. 245–252 (1988).

[2] ASTM C39, "Standard Test Method for Compressive Strength of Cylindrical Concrete Specimens," *Annual Book of ASTM Standards,* Vol. 04.02, pp. 19–21 (1988).

[3] ASTM C873, "Standard Test Method for Compressive Strength of Concrete Cylinders Cast in Place in Cylindrical Molds," *Annual Book of ASTM Standards,* Vol. 04.02, pp. 427–419, (1988).

[4] ASTM C116, "Standard Test Method for Compressive Strength of Concrete Using Portions of Beams Broken in Flexure," *Annual Book of ASTM Standards,* Vol. 04.02, pp. 55–56, (1988).

[5] ASTM C109, "Standard Test Method for Compressive Strength of Hydraulic Cement Mortars (Using 2-in or 50 mm Cube Specimens)," *Annual Book of ASTM Standards,* Vol. 04.01, pp. 74–79, (1986).

[6] Abrams, D. A., "Design of Concrete Mixtures," *Bulletin No. 1*, Standard Materials Research Laboratory, Lewis Institute, Chicago, IL. (1918).

[7] Lea, F. M., "The Chemistry of Cement and Concrete," Chemical Publishing Co., Inc., New York, NY, 3rd edition, pg. 392 (1971).

[8] Gruenwald, E., "Effect of Slump on Compressive Strength of Concrete Water/Cement Ratio," *A.C.I. Journal Proceedings*, Vol. 53, No. 2, pg. 230 (1956).

[9] ASTM C403, "Standard Test Method for Time of Setting of Concrete Mixtures by Penetration Resistance," *Annual Book of ASTM Standards*, Vol. 04.02, pp. 210–213 (1988).

[10] ASTM C191, "Standard Test Method for Time of Setting of Hydraulic Cement by Vicat Needle," *Annual Book of ASTM Standards*, Vol. 04.01, pp. 204–207 (1986).

[11] ASTM C187, "Standard Test Method for Normal Consistency of Hydraulic Cement," *Annual Book of ASTM Standards*, Vol. 04.01, pp. 192–194 (1986).

Chapter 3

WATER REDUCING CHEMICAL ADMIXTURES

INTRODUCTION

The author has chosen to discuss water reducing admixtures (WRAs) first because their volume of use in concrete is the largest of the chemical admixtures. This class of chemical admixtures permits the use of less water to obtain the same slump (a measure of consistency or workability), or the attainment of a higher slump, at a given water content, or the use of less portland cement to realize the same compressive strength. Their effects on the physical properties are specified in ASTM C494 [1]. The theoretical water-cement ratio ranges from 0.27 to 0.32, depending upon the composition of the portland cement and the individual doing the theoretical calculations. The amount of water in excess of this ratio is often called "water of convenience," in that it makes it more convenient to mix, transport, place, and finish the concrete.

MECHANISM OF WATER REDUCTION

When portland cement and water are mixed in a small container, using a metal or wooden spatula, tiny bubbles of water coated with cement particles can be observed, such as those illustrated in Figure 3-1. As mixing is continued, the tiny bubbles seem to disappear, at least to the naked eye, but they really never go away but only get smaller. This phenomenon also occurs in portland cement concrete, and the water encapsulated by the surrounding cement particles is not available for its workability, placing, and finishing. And, what is just as important, the surfaces of the cement particles which are abutting each other in the agglomerate are not available for early hydration, which represents approximately 12% to 20% of the cement surface.

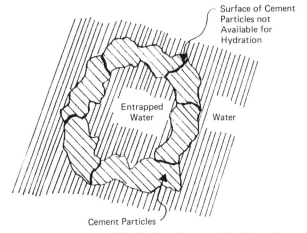

Figure 3-1. Cement particles in the absence of a dispersing agent.

When a water reducing admixture is added to concrete, the cement particles become dispersed in the aqueous phase of the concrete and do not form cement-water agglomerates. This results in (1) liberating the water that normally would have been entrapped by the surrounding cement particles so that it can contribute to the fluidity of the concrete and (2) making additional surfaces of the cement particles available for early hydration. A number of mechanisms have been proposed to account for the dispersive effect of the admixture. One of these postulates that the admixture is adsorbed on the surface of the cement particles and imparts like and repelling charges to the particles, as pictured in Figure 3-2. Not enough dispersant is present in the concrete mix, at the normally recommended addition rates, to completely coat the individual cement particles, but a sufficient amount is available to impart the like charge, whether it be positive or negative. The relative tendency of pure, synthetically prepared, phases of cement to adsorb a dispersant has been reported to be in the order

$$C_3A > C_4AF > C_3S = C_2S$$

but the work done with the pure phases was done in the absence of calcium sulfate and with only two dispersants—calcium lignosulfate and salicylic acid [2]. Another theory, which merits consideration, has to do with the adsorption of the disperant on the cement particles,

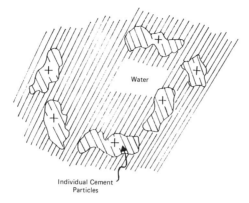

Figure 3-2. Effect of a dispersant on the cement-water mixture.

which through simple steric hindrance prevents one coated particle from approaching another, as shown in Figure 3-3. Micro photographs (500× magnification) of portland cement when mixed with water and with water containing a very small amount of a dispersing agent are shown in Figures 3-4*A* and 3-4*B*, respectively. The large clumps of cement particles become essentially individual particles in the presence of the dispersant.

COMPOSITION OF WRAs

That portion of the cement surface that has adsorbed the dispersant will be slower than normal to react with water (and with water and calcium sulfate, in the case of the C_3A and C_4AF phases) which tends

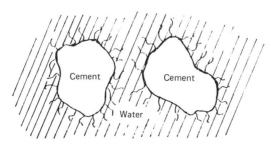

Figure 3-3. Dispersion through steric hinderance.

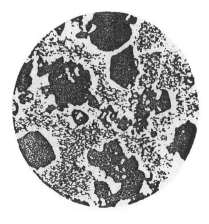 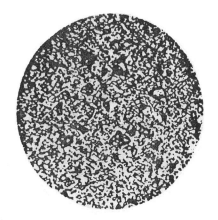

Figure 3-4*A*. Cement + water.   Figure 3-4*B*. Cement + water + dispersant.

to extend the time of setting of concrete and delay in strength development because it acts as a temporary barrier to the aggressive action of the water. It is for this specific reason that water reducing admixtures contain a second ingredient, which will accelerate the reaction of the exposed cement surfaces with water (and calcium sulfate) and off-set some of the delay in time of setting and aid in the development of early strength. The most commonly used second ingredients are chloride, nitrate, nitrite, formate, and thiocyanate salts of the alkali and alkaline earth metals, as well as triethanolamine. The function of these second ingredients and their relative performance as accelerators will be discussed in Chapter 4. It is recognized that there can be a delay in time of setting due to the addition of a water reducing admixture and ASTM has set a maximum of 90-minutes delay in both initial and final times of setting for a Type A admixture and has also set a minimum compressive strength gain at 3 days of 110% over that of the untreated, reference concrete. The most often used dispersing constituents of today's water reducing admixtures and their supplementary components are listed in Table 3-1.

HISTORY OF WRAs

The birth of water reducing admixtures occurred in 1930. At that time, many highways consisted of three lanes which led to many head-on collisions. The idea of paving the center lane with a dark colored

Table 3-1. Components and their Addition Rates of Modern Water Reducing Admixtures.

DISPERSANT NAME	ADDITION RATE[a]	SECOND INGREDIENT NAME	ADDITION RATE[a]
Alkali metal or alkaline earth salts of lignosulfonic acid	0.15	Triethanolamine Calcium chloride	0.01 0.30
Carbohydrates, such as glucose and corn syrup	0.04	Triethanolamine Calcium chloride	0.01 0.30
Alkali metal or alkaline earth salts of hydroxylated carboxylic acids (such as gluconic or heptogluconic acid)	0.06	Triethanolamine Calcium chloride	0.01 0.30

[a]% solids on weight of cement.

cement was conceived to warn the driver that he was in the hazardous lane. Extreme quantities of carbon black were needed to develop the dark color and a reduction of the treated concrete's strength, as well as its uniformity of color resulted. The answer to the problem of color was to be answered by the addition of a dispersant. Not only did the coloration of the concrete become remarkably uniform but when the concrete cores were taken from the center lane they were found to have substantially higher compressive strengths than those taken from the two outer lanes of traffic. It was concluded that the dispersing agent was not only acting on the carbon black but on the cement as well. The dispersant used was a sodium salt of a sulfonated formaldehyde-napthalene condensate, the forerunner of high range water reducing admixtures to be described and discussed later in this chapter.

The developers of the carbon and cement dispersant then began to search for other dispersants because of the cost of the napthalene-formaldehyde compound. Other investigators went on to develop a dispersing admixture consisting of calcium lignosulfonate, which was, and still is, a waste product [3][4]. The use of hydroxy alkyl amines, such as triethanolamine to counteract the set retardation caused by the dispersant was patented in 1936 [5]. In the years that followed, various combinations of salts of lignosulfonic acids and calcium chloride were found to produce good water reduction as well as a minimal retardation in time of setting accompanied by the early strength development. Although numerous formulations have been introduced to the concrete industry, those listed in Table 3-1 are the predominant ones.

The water reductions produced by three different commercially available water reducing admixtures in concrete (C.F. = 517 lbs) fabricated from a large number of different portland cements are summarized in Table 3-2. One important and evident conclusion that can be reached from a study of the data is that the degree of water reduction exhibited by the various portland cements varies widely (*note, the large standard deviations and coefficients of variation*). Although research is currently being conducted in an effort to establish the cause for this wide response to water reduction, the author believes that fineness, chemical composition of the clinker, and the nature of the calcium sulfate in the finished cement, or some combinations of these will end up being important factors (this will be discussed later in this chapter).

Another important, but not so evident, and to some quite unexpected, conclusion that can be drawn from the data in Table 3-2 is that the relationship of the concentration of the dispersing agent in the concrete to its response to water reduction is essentially linear. The author has taken the liberty of plotting the *average* water reduc-

Table 3-2. Water Reduction in Concretes Made from Different Portland Cements and Containing Different Dispersants.

TYPE OF DISPERSANT	AMT. OF DISPERSANT % SOLID ON WT. OF CEMENT	NO. OF CEMENTS TESTED	WATER REDUCTION		
			AVG. %	STD. DEV.	C.O.V. %
Alkali metal salt of a sulfonated naphthalene-formaldehyde condensate	0.375	95	12.0	4.32	35.96
Calcium lignosulfonate	0.147	77	4.1	2.84	68.37
Combination of calcium lignosulfonate and glucose polymer	0.069	63	2.6	2.35	90.50

tion in the many test concretes as a function of the amount of dispersant in the concrete systems in Figure 3-5. This tells one that (1) it is the addition rate of the dispersant that is important and (2) the chemical nature of the dispersing agent does not play a part in its *initial* water reduction. The word "initial" is purposely emphasized in the previous statement because the nature of the dispersant could well be an important factor in determining the retention (or loss) of slump in concrete with age, and this subject will be addressed later in this chapter.

ROLE OF CALCIUM SULFATE IN WATER REDUCTION

A part of the wide variability in response to dispersion by the cements in Table 3-2 is due to the form of calcium sulfate in the cement [6]. In 1962 the author first observed that concrete containing a water reducing admixture based on calcium lignosulfonate (CLS) being used in the construction of a runway at the Chicago O'Hare Airport was not responding to dispersion as expected and that it was undergoing very rapid stiffening. When the chemical admixture was omitted from the concrete mix design, the problem promptly disappeared. A similar problem had been experienced earlier by others [7]. This phenomenon

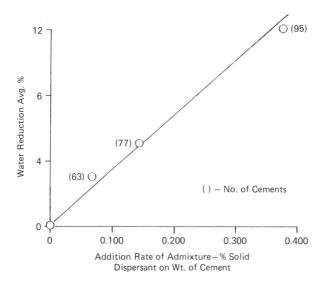

Figure 3-5. Influence of the addition rate of dispersant on the degree of water reduction.

has since been experienced in the U.S. on numerous occasions and has become known as "cement-admixture incompatibility." All of the cements exhibiting this behavior have one thing in common: they contain natural anhydrite in relatively large amounts.

Calcium sulfate, which is interground with portland cement clinker to control the hydration of the aluminate phases in the finished product, was briefly mentioned in Chapter 1. Calcium sulfate is added to the finish grinding mill either as gypsum or a mixture of gypsum and natural anhydrite. Both gypsum and natural anhydrite occur in nature, often in adjacent deposits, and differ in the amounts of water bound in their crystal structures. Theoretically, the chemical formula for gypsum is $CaSO_4 \cdot 2H_2O$ and that of natural anhydrite is $CaSO_4$ (however, the latter usually contains traces of combined water). But what is more important, the two forms of calcium sulfate differ in their rate of solution in the aqueous phase of a cement-water mixture.

Samples of gypsum and natural anhydrite were ground separately, at room temperature, to a Blaine fineness of $3000 \pm 20 \ cm^2/g$, in laboratory steel ball mills. Then a portland cement clinker, typical of that used to produce a Type I cement, was ground to approximately the same fineness and then blended with the ground gypsum, in one case, and with the ground natural anhydrite in the second instance. Both blends were designed to have the same sulfate ion content; i.e., $3.00\% \ SO_4^=$ (or $2.50\% \ SO_3$). Samples of the two clinker-calcium sulfate blends were then examined for their early stiffening properties [8].

The test data, illustrated in Figure 3-6, indicate that the cement containing natural anhydrite exhibited early stiffening which could only be partially destroyed by remixing after 11 minutes, and stiffening continued until the end of the test period. The rate of stiffening of the mortar made from the cement containing gypsum was fairly normal. When CLS is added to the mortars (0.125% solids on weight of cement), the shape of the penetration vs. time curve for the cement containing the natural anhydrite undergoes a drastic change, while there is very little change in the rate of stiffening profile of the cement containing gypsum (Figure 3-7).

The reason for the difference in stiffening characteristics, illustrated in Figures 3-6 and 3-7, becomes apparent when the solubility of the sulfate ion component of the two cements is determined, with and without CLS being present, at a W/C of 0.75. The data are exhibited

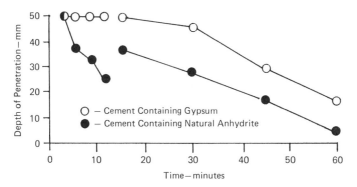

Figure 3-6. Results of modified ASTM C359 test procedure.

in Figure 3-8. It is the author's opinion that the repression of the solubility of the natural anhydrite is the result of the CLS being adsorbed on its surface. This results in a reduction of its sulfate supply rate and leads to the aluminates undergoing direct reaction with water, resulting in rapid stiffening.

Up until about 1949, it was generally believed that natural anhydrite could not be used to control the rate of the hydration of the aluminates because its rate of solution in the aqueous phase of concrete was too slow to prevent early stiffening. It then was reported that mixtures of natural anhydrite and gypsum could be used to replace gypsum without adversely affecting either the time of setting of cement paste or the strength and volume change characteristics of concrete; and in some cases it was found possible to completely sub-

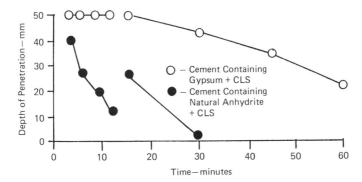

Figure 3-7. Results of modified ASTM C359 test procedure in which CLS was present.

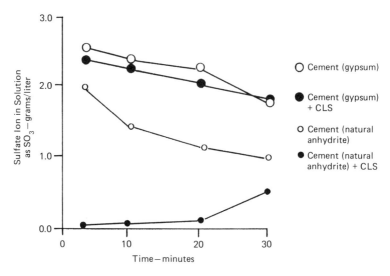

Figure 3-8. Sulfate solubilities of Portland Cement made from gypsum and natural anhydrite, with and without CLS.

stitute natural anhydrite for gypsum [9]. One other point should be made here. The work was done in the absence of water reducing admixtures, because they were in their infancy.

The author's experimental results indicate that natural anhydrite, whether interground purposely or as an impurity in gypsum, with portland cement clinker can produce rapid set and reduce the expected water reduction when the common water reducing admixtures are added to the concrete. This situation is not limited to the addition of CLS. Water reducing admixtures based on salts of hydroxylated carboxylic acids as well as carbohydrates (sugars, corn syrup, etc.) act very much like the CLS type when added to concrete containing portland cement, wherein the sulfate ion contributor is natural anhydrite. The author's most recent research results indicate that when the cement contains a ratio of gypsum to natural anhydrite (by weight) of less than 2, rapid stiffening and very poor response to dispersion (water reduction) will occur when these basic chemical water reducing ingredients are added.

INFLUENCE OF DISPERSION ON PROPERTIES OF CONCRETE

The author pointed out earlier that the strength contribution of portland cement to concrete is enhanced in the presence of a water reducing admixture through (1) the lowering of the W/C and (2) increasing the surface area of the cement particles for early hydration. By employing the O.I.F. concept (Chapter 2), the effect of this increase in cement surface area can be isolated and calculated. The six concretes used to illustrate this are described in Table 3-3. Three of the concretes were treated with three different, but commonly used, water reducing admixtures. The compressive strengths at three ages of test of the six concretes are plotted as a function of the O.I.F.s in Figures 3-9 (1 day), 3–10 (7 days), and 3–11 (28 days). If the water reducing admixtures functioned only through a W/C reduction, their data points would fall on the reference base lines in each of the three figures because W/C is accounted for on the abscissa. All of the data points for the admixture treated concretes fall above the reference baseline by a significant margin. The difference in psi between those data points and those on the reference base line, having the same O.I.F., represent the increase in compressive strength provided by the additional cement surfaces made available for hydration through dispersion. If the admixture does something chemically to alter the rate of the cement hydration or the nature of its hydration products, this effect will also be a portion of the psi difference. The strength increases realized through the use of the three are summarized in Table 3-4.

Table 3-3. Concretes Containing Three Common Chemical Water Reducing Admixtures.

COMPOSITION[a] AND PHYSICAL PROPERTIES	CONCRETE NO.					
	1	2	3	4[b]	5[c]	6[d]
Cement—lb/yd³	550	521	470	500	500	505
Coarse Agg.—lb/yd³	1635	1645	1675	1685	1680	1675
Fine Agg.—lb/yd³	1340	1345	1370	1380	1375	1370
Water—lb/yd³	300	307	300	280	280	285
W/C	0.54	0.59	0.64	0.56	0.56	0.56
O.I.F.	1019	883	784	893	893	902
Air—%	1.8	1.6	1.6	1.7	2.0	1.8
Slump—in.	3-1/2	3-1/4	3-1/4	3	3-1/2	3-1/4
Comp. Strength—psi						
1-day	1490	1275	1110	1475	1490	1460
7-days	4110	3540	3125	4080	4090	4175
28-days	6500	5635	4990	6385	6250	6440

[a]Amounts adjusted to 27 ft³/yd³ yield.
[b]Admixture based on a salt of a hydroxylated carboxylic acid.
[c]Admixture based on CLS.
[d]Admixture based on corn syrup + CLS.

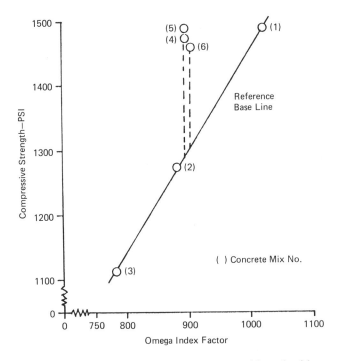

Figure 3-9. 1-day compressive strength of concretes, with and without water reducing admixtures, as a function of their O.I.F.s.

In addition to their influence on strength when their application is focused on W/C reduction, water reducing admixtures exert the following other effects on the properties of concrete.

1. Because of the W/C reduction, the shrinkage of the treated concrete is reduced, mainly because of a reduction in drying shrinkage or possibly a change in the structure or composition of the calcium silicate hydrates. Some might argue this point, but the method of measurement leaves much to be desired because of its precision [10]. It is true that when the set accelerating component of the water reducing admixture is used by itself as an admixture and in considerably large amounts, more so than those employed when it is introduced as a component of water reducing admixtures, it can cause an increase in shrinkage, but in these cases a water reduction is not realized and most of the shrinkage is due to chemical reasons.

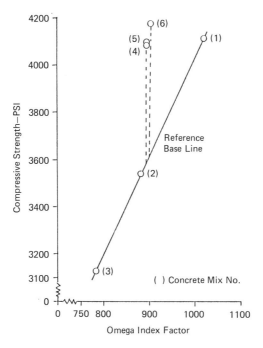

Figure 3-10. 7-day compressive strengths of concretes, with and without water reducing admixtures, as a function of their O.I.F.s.

2. The reduction in W/C also densifies the concrete (of all of its components, water has the lowest specific gravity) making it less permeable to aggressive water and its salt solutions. This means that the treated concrete is less susceptible to chloride and sulfate ion intrusion. The former can aggravate the corrosion of embedded steel (Chapter 4), and the latter can give rise to sulfate attack (Chapter 7).

3. The dispersant present in the water reducing admixture can increase the amount of air entrained and/or trapped in the concrete, whether an air entraining admixture is absent or present. While the dispersant is basically a solid-water dispersant, it can to a minor extent, disperse air in water. For example, a plain concrete that has a 1.5% content, will have its air content increased to about 2.5% when a water reducing admixture based on calcium lignosulfonate is added; while those based on salts of hydroxylated carboxylic acids, and on glucose polymers only increase the air content to 1.8%.

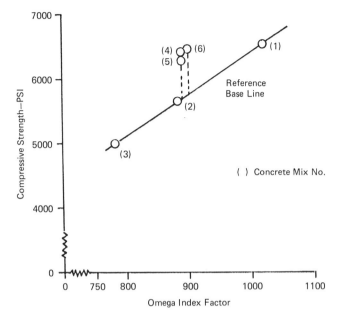

Figure 3-11. 28-day compressive strengths of concretes, with and without water reducing admixtures, as a function of their O.I.F.s.

Table 3-4. Influence of Water Reducing Admixtures on the Compressive Strength of Concrete Determined by the Omega Index Factor Concept.

CONCRETE NO. (TABLE 3-3)	4	5	6
1-day compressive strength—psi	1475	1490	1460
ΔPSI Actual data point-data point on reference base line (same O.I.F.)	180	195	150
P.S.I./lb cement/yd³ concrete increase	0.4	0.4	0.3
Increase over reference point—%	15	15	11
7-day compressive strength—psi	4080	4090	4175
ΔPSI-Actual data point-data point on reference base line (same O.I.F.)	490	500	555
PSI/lb cement/yd³ concrete increase	1.0	1.0	1.1
Increase over reference point—%	14	14	14
28-day compressive strength—psi	6385	6250	6440
Δ PSI Actual data point-data point on reference base line (same O.I.F.)	685	550	690
PSI/lb cement/yd³ concrete increase	1.4	1.1	1.4
Increase over reference point—%	12	10	12

EFFECT OF CEMENT ALKALI AND ALUMINATE PHASES ON DISPERSION

One of the observations the author has made, and he is sure that others have too, is that water reduction is more noticeable in concretes made from cements of low alkali and/or C_3A contents and in lean mixes. The alkali, C_3A and C.F. relationships are illustrated in Figures 3-12, 3-13, and 3-14, respectively. The water reducing admixture used to produce the data in these three figures was based on a combination of calcium lignosulfonate and triethanolamine at an addition rate of 0.18% solids on weight of content. The cements chosen for the work shown in Figure 3-12 had a total alkali content (as Na_2O) in the range

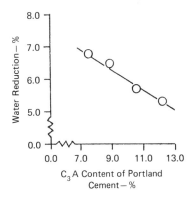

Figure 3-12. Influence of C_3A content on water reduction.

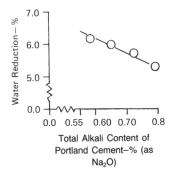

Figure 3-13. Influence of alkali content on water reduction.

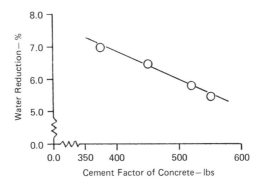

Figure 3-14. Influence of cement factor on water reduction.

of 0.63% to 0.67% and a fineness that varied from 3550 to 3610 cm^2/g, and the concretes had a C.F. of 517 lb. The data in Figure 3-13 were generated from concretes whose C.F. was 517 lb and contained cements whose C$_3$A contents varied over a fairly narrow range, i.e., 9.0% to 9.5%, and their fineness values were limited to a range of 3340 to 3380 cm^2/g. The concretes described in Figure 3-14 were fabricated from a cement whose total alkali (as Na$_2$O) and C$_3$A contents were 0.67% and 10.3%, respectively, and whose fineness was 3420 cm^2/g.

With respect to Figure 3-12, the author mentioned earlier that there is good evidence that the C$_3$A and C$_4$AF phases of portland cement were strong adsorbers of dispersants in the absence of sufficient sulfate ions to protect their surfaces with ettringite and/or ettringite-like compounds [2]. The SO$_3$ contents of the cements used in Figure 3-12 varied, and the source of sulfate ions, whether it be gypsum, hemihydrate, etc., was not determined. Regardless of the source of the SO$_3$, "Dodson's Rule No. 10" states that the ratio of the percent of C$_3$A to the percent of SO$_3$ (as reported in the cement mill certificate) must be less than 2.5, and preferably close to 2.0, in order to realize the maximum performance of chemical admixtures, whether they be water reducing, set retarding, or set accelerating admixtures.

In regard to Figure 3-13, the result of reducing the total alkali content of the cements used to fabricate the test concretes is probably due to the alkali-sulfate relationship, which results in a lower solubility of the sulfate ion in the concrete-water phase and allows a more rapid adsorption of the dispersant by the aluminate phases. The data of Kalousek, who analyzed the aqueous phase of a number of different portland cement-water mixtures clearly illustrates this point, a part of which is illustrated in Figure 3-15 [11]. In that work, the SO_3 was added to each cement clinker in the form of finely ground gypsum so that all of the cements contained 1.75% SO_3. Each of the cements was mixed with water at a water-cement ratio of 0.35, and the liquid phase was removed by filtration and analyzed. In one series of experiments a mixing period of seven minutes was employed, and in the second series the filtration was done after two hours of mixing. The two sets of data points in Figure 3-15 fall on a relatively smooth curve. The SO_3 content of the aqueous phase, when the soluble alkali content is close to a value of zero, approximates the value of 1.09 g/L, its solubility in saturated lime water. The 0.100 moles/L shown in Figure 3-15 represents a 0.32% soluble alkali in the cement.

DELAYED ADDITION OF WRAs

The question of who and what adsorbs the dispersant in portland cement that is properly retarded by calcium sulfate has bothered the author for some time. It has been shown by others that the aluminate

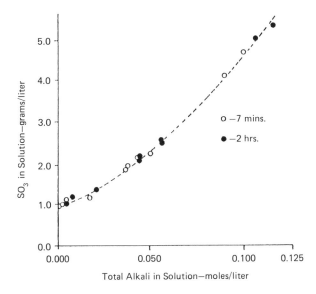

Figure 3-15. SO_3 solubility in cement-water aqueous phase vs. soluble alkali.

phases, either in the presence or absence of calcium sulfate, are strong adsorbers of the dispersant, but the adsorbancy of the silicate phases have been either ignored or assigned a low level [2][12][13][14]. The consequences of adding the dispersant to the concrete (or mortar) after a period of mixing—or what has come to be known as "delayed addition"—can be considered as a possible answer [15][16].

When a dispersant is added to concrete several minutes, or even seconds, after all of the concrete ingredients have undergone agitation, a number of changes in the properties of the concrete are observed when compared to those of the same concrete in which the dispersant is added with the mix water. For example, at a given W/C, the slump is increased, the amount of air entrained by the dispersant is increased, and the initial and final times of setting are increased. These observations are interpreted by the author as follows.

1. When the dispersant is added with the mix water, a great portion of it is adsorbed by the aluminate phases before its reaction products with calcium sulfate are fully formed, and very little of it is adsorbed by the silicate phases [2].
2. After a brief period of mixing, the calcium alumino sulfate compounds have started to form, and these are less prone to adsorb the dispersant when its addition is delayed. This results in having more dispersant free to either entrain air or be adsorbed by the calcium silicate hydrate compounds.
3. Since the hydration of the C_3S phase is a major contributor to the time of setting, the interference of its natural hydration by the adsorbent delays the time of setting.

HIGH RANGE WATER REDUCING ADMIXTURES (HRWRs)

The newest water reducing admixtures are those of the high range type which are often called super water reducers or "supers." They will simply be referred to as HRWRs (High Range Water Reducing Admixtures). They are described as the "newest," but they were really the long forgotten embryo of the common water reducing admixtures, described earlier in this chapter. This type of admixture provides a far greater degree of water reduction in concrete, through dispersion, without causing excessive retardation of time of setting and without entraining large amounts of air. The author believes that they function through adsorption by the cement phases and provide dispersion through steric hindrance but offer only a minimum resistance of the cement surfaces to chemical attack by water.

ASTM has established stringent performance requirements for this kind of chemical admixture; i.e., type F water reducing, high range [1]. These include a minimum of 12% water reduction (vs. a minimum of 5% for a type A water reducing admixture) and a minimum of 140% increase (vs. the reference, plain concrete) in 1-day compressive strength. The allowable effect of the type F admixture on the time of setting of concrete is the same as that set forth for the type A. The basic reason for their greater dispersive power lies in their addition rates; that of the type F is usually 3 to 7 times higher than that of a type A (Figure 3-5). This is the main reason the author favors the steric hindrance mechanism through which they act as dispersants. More will be said about this later in this chapter when the subject of slump loss is broached.

The HRWRs are based essentially on two complex organic materials: (1) salts of sulfonated melamine-formaldehyde condensates and (2) salts of sulfonated naphthalene-formaldehyde condensates. The chemical structures proposed for these two products are illustrated in Figures 3-16 and 3-17. For the sake of brevity, the first is refered to as HR-M (M = melamine) and the second as HR-N (N = naphthalene). On rare occasions, large dosage rates of purified salts of lignosulfonic acid are used but, because of the severe delay in the time of setting of the treated concrete, they are usually accompanied by a large dosage of a set accelerating admixture (Chapter 4). As men-

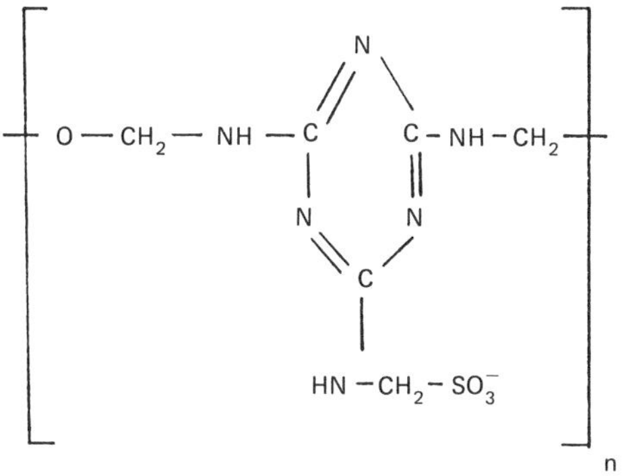

Figure 3-16. Sulfonated melamine-formaldehyde condensation product.

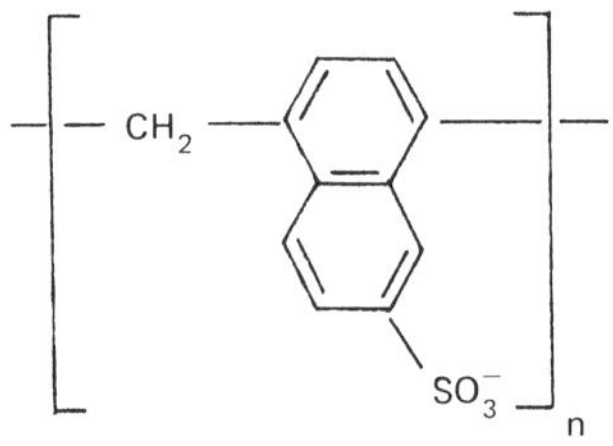

Figure 3-17. Sulfonated naphthalene-formaldehyde condensation product.

tioned earlier, HR-N was found to be an excellent dispersant for portland cement in concrete as early as 1930 [17]. The date at which the
HR-M type was first observed to have dispersing powers for portland
cement cannot be reliably established, but its principal developmental
work was done in Europe during the mid-1960s.

When the "supers" were first introduced to the U.S. concrete market, it proved to be a "tough sell" situation for the admixture producers because of their high use rate-cost factors, uncertainties about
their long-term effect on the properties of concrete, and the absence
of ASTM specifications covering their performance requirements. Their
rather sluggish start has taken a 180° turn during the past 15 years
because of the following factors:

1. Reduction in their use rate-cost factors.
2. Development of convincing evidence that they do not create long-
   term problems in concrete.
3. Establishment of performance requirements by ASTM.
4. Rapid growth of the demand for either "flowing" concrete and/
   or high strength concrete by the construction industry.

The rapid rise in the demand for "flowing" concrete has come about
as a result of the soaring cost of construction labor. The addition of
the "supers" can increase the slump (or workability, or placeability)
by a factor of three without the addition of more water (which would
increase the W/C), and the contractor finds his labor costs involved
in the placement of the concrete to be much lower, especially when
the forms contain a large amount of metal reinforcement. Because of
the ease in placement, more concrete can be handled, per unit time,
and the total job time as well as the number of laborers needed are

reduced. Of course, the contractor could attain his "flowing" concrete by simply specifying that more water be added to the concrete, either at the ready mix plant or at the job site. However, this approach to "flowing" concrete has numerous negative effects on the quality of the concrete. Among these are segregation, excessive bleeding, loss of entrained air (if entrained air is specified), and a reduction in compressive strength because of the increase in W/C. The concrete producer also finds that the use of "flowing" concrete is to his/her economic advantage because delivery trucks spend less time at the job site dispensing the concrete into its forms; and, because of its flowing characteristics, the amount of concrete normally retained in the mixing drum of the truck is minimized, which reduces the waste concrete disposal problem that today is facing every producer.

An example of "flowing" concrete produced in the author's laboratory, through the use of an HRWR, is illustrated in Table 3-5. At least the following three factors should be noted in that data:

1. The fine aggregate in the "flowing" concrete was increased by approximately 8% to improve the cohesiveness of the very fluid mix, and the coarse aggregate was slightly reduced.
2. The increase in compressive strength of the treated concrete, at

Table 3-5. Properties of Flowing Concrete.

CONCRETE COMPONENT	REFERENCE	FLOWING	COMPENSATED
Cement—lb/yd^3	600	600	450
Fine Agg.—lb/yd^3	1200	1300	1425
Coarse Agg.—lb/yd^3	1900	1800	2000
Water—lb/yd^3	300	300	225
HR-N—% solids on cement	—	0.50	0.50
W/C	0.50	0.50	0.50
Physical Properties			
Slump—in.	3-1/2	9-1/2	5
Air—%	1.7	1.9	1.6
Compressive Strength—psi			
1-day	1410	1720 (122)[a]	1520 (108)[a]
2-days	2750	3170 (115)	2900 (105)
7-days	4100	4530 (110)	4280 (104)
28-days	5120	5560 (109)	5330 (104)

[a]( ) % increase over the reference, which is 100%.

all ages of test, was not due to a reduction in W/C or an increase in coarse aggregate contents.

3. The slump of the concrete treated with the admixture was increased 6 inches (over that of the reference).

Now, consider the role that HRWRs can play in the production of high strength concrete. What is high strength concrete? When asked this question, the answer is that it is concrete designed at a minimum water-cement ratio to obtain the highest ultimate strength. While this answer is rather vague, the term, high strength, is also ambiguous, so the author leaves it up to the reader to answer the question. One way to attain this goal, whatever it is, is through the use of HRWRs; another way will be discussed in Chapter 7, wherein the combination of the HRWRs with pozzolanic mineral admixtures is considered.

ECONOMICS IN USE OF HRWRs

There are many ways in which the economics of using high strength concrete express themselves. In the case of prestressed concrete, there can be a reduction, or even elimination, of steam curing, which represents a considerable savings in energy costs. In addition, a type I (or type II) portland cement rather than the more costly type III cement can be used. Probably the most dramatic savings come from the reduction in form cycle time and an increase in production because of the more rapid increase in early release strength.

Next, consider conventional structural concrete, with or without steel reinforcement, wherein the early stripping of forms also becomes an economical factor. The shorter the time period between the start of construction and the occupation of the building by paying tenants, the greater are the savings to the developer. Higher compressive strengths allow the engineer to design smaller size supporting members to carry the same loads as a large number made of concrete of ordinary strength. For example, when elements exposed to compressive forces, such as columns and core walls were constructed from high strength concrete (by changing the specified compressive strength from 5800 to 8700 p.s.i.) in one multi-story construction, the increase in each floor's rental space resulted in an effective benefit of approximately $100,000 rental per floor for the client [18].

One of the conclusions reached in a study of steel reinforcement column costs was that more economy could be realized through the use of high strength concrete than with high strength reinforcing steel [19]. In another study, it was calculated that by increasing the compressive strength from 4000 to 8000 psi, for a 15 story structure, the cost of the concrete in the supporting columns could be reduced by 21% to 43%, depending upon the size of the column and the span [20]. Further savings in the use of high strength concrete in columns are realized by permitting the designer to maintain a constant column dimension throughout several stories of a multi-level structure, which results in an increase in the use factor of the forms.

Another economic aspect frequently considered by the concrete producer in using HRWRs is the potential capability, and savings, of using less cement in the concrete to produce the same specified compressive strength as normal, admixture-free, concrete. Many refer to this as "compensated" concrete and an example is illustrated in the right hand column in Table 3-5. Some of the savings realized by the reduction in the cement factor is offset by the cost of additional fine and coarse aggregates (to maintain yield) as well as that of the admixture. Although there might be some net savings to gain, the author is not a strong advocate of this approach.

Adding water to concrete to increase its workability, rather than a water reducing admixture, can cost the contractor money. Its addition increases the W/C of the concrete, which, in turn, decreases its compressive strength. This means that the C.F. will have to be increased to attain specified strength levels. The cost of increasing the water content from 4 to 6 gal/yd^3 was estimated, almost 10 years ago, to be close to $2.00 [21]. The author's cost calculations, based on today's prices, raises that value to about $2.25.

When it comes to concrete, the author considers excess water as a very uncooperative fluid. In addition to increasing the W/C, some of the excess water needed for workability either bleeds to the surface or becomes entrapped on the underside of the coarse aggregate particles, as illustrated in Figure 3-18. In the first case, the excess water on the surface reduces the abrasion resistance of the surface. If the entrapped water is not able to channel its way to the surface, it becomes a point of deposition for crystalline calcium hydroxide produced by the hydrating cement, and the paste and/or mortar does not have an opportunity to form a strong bond to the under-surface of the aggregate. If the water originally collected below the coarse aggregate

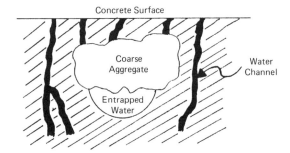

Figure 3-18. One of the results of excess water in concrete.

does manage to channel to the surface, an empty void in the concrete mass results. Either way, the integrity of the concrete is disrupted.

It should be pointed out that an increase in the slump of plain concrete by 1 inch represents an increase of about 0.015 in W/C. This is an average of the author's observations made on a large number of concretes fabricated from different portland cements over a wide range of C.F.s. Therefore, increasing the slump of plain concrete from 3″ to 7″ (W/C = 0.55), by the addition of water, results in an increase of W/C of approximately 11%.

TIME OF SETTING OF HRWR CONCRETE

There are a number of things that bother the author about the influence of HRWRs on the behavior of the plastic concrete (that which has not yet undergone initial set). For example, it is difficult to explain why, at the addition rates commonly used, they do not delay the time of setting of concrete to an alarming extent. Typical time of setting data are shown in Table 3-6. The air entraining depressant (sometimes called an air detrainer), tributyl phosphate, used in concrete No. 4 (Table 3-6), was added to bring its air content down to a level comparable to that of the other concretes, because CLS tends to entrain air. The retardation of time of setting by the CLS, when added at the same dosage rate as the HR-N and HR-M is 5 to 6 times that of the HRWRs. Yet, at the same addition rate, the water reduction (as indicated by the slump of the concrete) realized through the use of the CLS is essentially the same as that created by HR-N and HR-M. In the case of concrete No. 6, where the HR-N and a rela-

Table 3-6. Properties of Concretes Treated with Various Water Reducing Admixtures.

CONC. NO.[a]	ADMIXTURE	ADDT'N RATE OF ADMIXTURE[b]	W/C	AIR CONTENT-%	SLUMP-IN	TIME OF SET-MIN INITIAL	FINAL
1	none	—	0.55	1.8	3	281	382
2	HR-N	0.375	0.45	2.2	8-1/2	355	435
3	HR-M	0.375	0.45	2.3	8-1/4	350	440
4	CLS+ AED[c]	0.375 —	0.45	2.0	8	625	770
5	CLS	0.200	0.50	2.5	5	430	532
6	HR-N+ CLS	0.375 0.150	0.42	2.3	9	390	475

[a]C.F. = 517 lb, type I cement.
[b]% solids on weight of cement.
[c]AED = air entraining depressant.

tively small amount of CLS were added, the initial slump increased by 1/2" and the initial and final times of setting were extended 35 to 40 minutes, respectively, over those of the concrete (No. 2) that contained only the HR-N. This is one of the reasons that leads the author to believe that the CLS and the HRWRs act in different ways to disperse the cement particles and that the latter behaves as it does through steric hindrance.

Another way to view the dispersion-set retardation relationship is through the composition of the HRWRs. The proposed molecular structures for HR-M and HR-N are illustrated in Figures 3-16 and 3-17. The structure of CLS is not well defined, but its total sulfur content can be determined and then stoichiometrically converted to the number of *sulfonate* groups, $R-SO_3^- \, M^+$, where R is an organic constituent, M^+ a monovalent cation and SO_3^- is the sulfonate group. The sulfonate group content of the three admixtures is listed in Table 3-7. The values cited for CLS are probably somewhat higher than

Table 3-7. Active SO_3 Groups in Water Reducing Admixtures.

ADMIXTURE	SO_3 GROUPS—%
HR-N	36–37
HR-M	30–31
CLS	21–22

actual because of the inorganic ($CaSO_3$) sulfite contamination resulting from the sulfite pulping process, but the amount of active sulfonate groups in the CLS is definitely lower than that of the other
two admixtures. Therefore, at the same addition rate, the HRWRs
would provide between 60% and 70% more sulfonate groups than
CLS. If the sulfonate group acts like the sulfate ion (although having
a single negative charge), it may well readily react with the tricalcium
aluminate phase of the portland cement and prevent its direct reaction
with water much like the sulfate ion (from calcium sulfate) and provide dispersion. Others have found that portland cements containing
gypsum respond better to the addition of HRWRs than those in which
a part or all of the calcium sulfate is present as hemi-hydrate, and
exhibit less slump loss [22]. This falls into a reasonable pattern in
that the hemi-hydrate (because it supersaturates the aqueous phase of
the concrete) would offer greater competition between the sulfonate
ion which would offer dispersion, and the sulfate ion, which provides
no dispersion, with the tricalcium aluminate phase of the cement.

The data presented in Figures 3-12, 3-13, and 3-15 indicate the
following:

$$\text{water reduction} \simeq \frac{1}{\text{alkali content of cement}}$$

$$\simeq \frac{1}{C_3A \text{ content of cement}}$$

$$\simeq \frac{1}{SO_3 \text{ content of cement}}$$

Putting all of these factors together, the degree of water reduction is
inversely related to the C_3A content of a given portland cement because the alkali-SO_3 factors cancel out in the overall relationship.

LOSS OF SLUMP OF PLAIN CONCRETE

Another thing that bothers the author about the HR-N and HR-M type
of HRWRs is their abnormal effect on the rheology of fresh concrete,
often expressed as slump. Slump is generally agreed to be a measure
of the consistency of concrete, as measured by the standard slump
cone [23]. It is also a measure of the workability of concrete and

determines the ease and homogeneity, with which it can be mixed, placed, compacted, and finished. The abnormal change in rheology just mentioned has to do with the treated concrete's rapid loss of slump. Slump loss is the amount by which the slump of freshly mixed concrete changes during a period of time after the initial slump measurement.

All concretes, either plain or treated with a chemical admixture, undergo some degree of slump loss. While many explanations have been proposed to account for this behavior in plain concrete, it is the result of three on-going processes. *First,* the tricalcium aluminate phase of the cement reacts very rapidly with water (and calcium sulfate) to tie up to 5% to 10% of the water originally added to the concrete, depending upon the C.F., W/C, and the composition of the cement. This reduction in available water for fluidity will reduce its slump. *Second,* the dormant period of the tricalcium silicate phase in cement pastes at 72° F, unagitated and immersed in a heat sink, can extend for various periods of time, depending upon the water-C_3S ratio. This is illustrated by the isothermal conduction curves in Figure 3-19. Neither the concrete central mixer nor the ready mix truck are isothermal, so most of the heat produced by the mixing action and the hydration of the cement is retained within the cementitious mass. *Third,* during the mixing operation, the coarse and fine aggregates are constantly abrading the cement particles and as a result, the coating of hydration

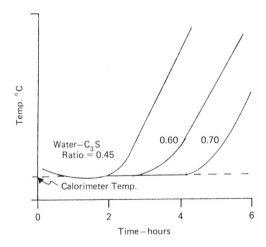

Figure 3-19. Influence of water-C_3S ratio on the dormant period of synthetic C_3S.

products that is envisioned as forming on the aluminates and silicates to produce the dormant period has its chances of protecting further hydration considerably reduced. Not only does the early hydration of the C_3A and C_3S phases tend to remove water from the mass of the concrete mix, but their hydration products act to stiffen the mix. The abrasion effect created during the mixing of the concrete is further manifested in the fact that the greater its initial slump (and fluidity) the greater is its slump loss—at least over a 90-minute period of mixing. The temperature of the concrete is also important. As the temperature is increased, its loss in slump, regardless of its initial slump, is increased.

The author's laboratory slump data generated from concretes in which the raw materials and mixer were pre-conditioned, in one case at 72° F and in the second case at 85° F (and the test concretes were batched at those two temperatures) are shown in Figure 3-20. The magnitude of each slump value was recorded only to the nearest $1/4''$, because of the lack of accuracy of the method of measurement. The data illustrated in Figure 3-20 indicate that:

- For a given cement, the slump loss, over a period of 90 minutes of mixing, increases with an increase in concrete temperature.

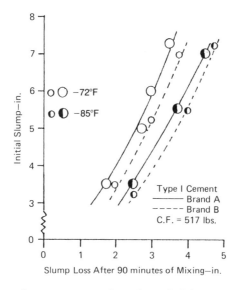

Figure 3-20. Effect of temperature on slump loss of plain concrete vs. initial slump.

- The slump loss, at a given temperature, can vary depending upon the cement, even though the cements may be of the same type.
- As the initial slump of the concrete, regardless of its temperature, is increased, its loss in slump is increased.

Although the position of the curves in Figure 3-20 for the two different brands of a type I cement are not super-imposable at either temperature of test, they have essentially the same configuration.

LOSS OF SLUMP OF HRWR CONCRETE

Thus far the author has confined his remarks to the loss of slump of plain concrete. The author's second concern about HRWRs, i.e. their effect on the slump of concrete over an extended time of mixing now must be addressed. For example, compare the slumps of the six concretes described in Table 3-6 as a function of time of mixing as they are shown in Figure 3-21. The data illustrated in Figure 3-21 indicate that after 90 minutes of mixing:

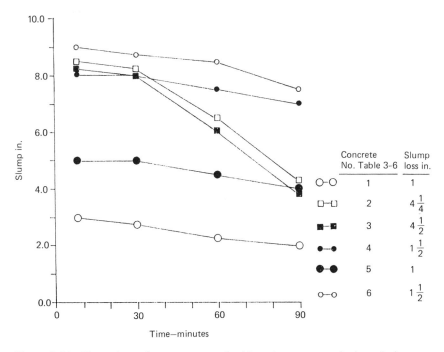

Figure 3-21. Slump loss of concrete treated with various water reducing admixtures.

- The concrete treated with CLS (No. 5) at an addition rate that would qualify it as a type D admixture (see Chapter 5) exhibits the same loss in slump as the reference (Concrete No. 1).
- The concrete treated with CLS (No. 4) at the same addition rate as that for the two HRWRs has essentially the same initial slump as those of the HR-M and HR-N treated concretes, *but* the magnitude of its slump loss is much less than the latter two and is the same as that of Concretes No. 1 and 5.
- By adding the CLS at a reduced dosage rate, along with the HR-N (Concrete No. 6), the slump loss is markedly reduced and the time of setting is only increased by 35 to 40 minutes over that of the HR-N treated concrete (No. 2).

One other characteristic difficult to explain is the concrete's response to change in slump by the addition of a HRWR in several increments or as a single dose to concrete after various periods of mixing. Delayed addition of conventional water reducing admixtures was mentioned earlier in this chapter, but extended periods of delayed addition were not investigated in that work [16]. The delayed addition of HR-N (as well as HR-M, but not shown) results in a rather peculiar slump behavior, as illustrated in Figure 3-22. This phenomenon has been confirmed, to some degree, by other investigators [24].

The data in Figure 3-22 warrant some comment. The "saw-tooth" configuration of the plot for Concrete No. 1, which was treated with four successive additions of HR-N (each at 0.4% solids on weight of cement) at 0, 60, 90, and 120 minutes underwent a slump loss of 2″ (8-1/4″ to 6-1/4″). Concrete No. 2, which was treated with two additions of HR-N (both at 0.4%, originally and after 120 minutes of mixing) exhibited a 3″ slump loss. Concrete No. 3, which was not originally treated with HR-N but dosed after 120 minutes of mixing with 0.4% HR-N, showed a gain in slump, after the addition, of 3-1/4″ (4-3/4″ to 1-1/2″). And finally, test Concrete No. 4 (whose slump data are not shown in Figure 3-22 because of its already complicated nature) had an initial slump of 3-1/4″ with no admixture added. At the end of the 120 minute mixing period its slump had dropped to 1-3/4″. After the addition of the HR-N, at a rate of 0.80%, its slump increased by 5-1/4″.

The author's only explanation, at this point in time, for the changes in slump just cited for the four concretes is two-fold:

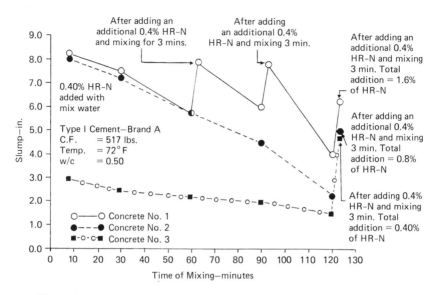

Figure 3-22. Effect of delayed addition of HR-N on slump of concrete.

1. The dispersant becomes engulfed by the cement hydration products and can no longer act as such which creates a rapid loss in slump.
2. The hydration product-dispersant coating is sloughed off by the abrading mixing action, and when the addition of the HRWR is delayed it is adsorbed by the freshly exposed cement surfaces to produce a sudden surge in dispersion and hence a marked increase in slump.

REFERENCES

[1] ASTM C494, "Standard Specification for Chemical Admixtures for Concrete," *Annual Book of ASTM Standards,* Vol. 04.02, pp. 245–252 (1988).
[2] Blank, B., Rossington, D. R., Weinland, L. A., "Adsorption of Admixtures on Portland Cement," *Journal of the American Ceramic Society,* Vol. 46, No. 8, pp. 395–399 (1963).
[3] Mark, J. G., "Concrete and Hydraulic Cement," *U. S. Patent, No. 2,141,570,* Dec. 17 (1938).
[4] Scripture, E. W., "Cement Mix," *U. S. Patent, No. 2,169,980,* Aug. 15 (1939).

[5] Tucker, G. R., "Amine Salts of Aromatic Sulfonic Acids," *U. S. Patent, No. 2,052,586,* Sept. (1936).

[6] Dodson, V. H., Hayden, T. D., "Another look at the Portland Cement/Chemical Admixture Incompatibility Problem," *Cement, Concrete, and Aggregates,* CCAGDP, Vol. 11, No. 1, pp. 52–56, Summer (1989).

[7] Manabe, T., Kawada, N., "Abnormal Setting of Cement Paste Owing to Calcium Lignosulfonate," *Semento Konkurito,* No. 162, pp. 24–27 (1960).

[8] ASTM C359, "Standard Test Method for Early Stiffening of Portland Cement (Mortar Method)," *Annual Book of ASTM Standards,* Vol. 04.01, pp. 270–273 (1986).

[9] Hansen, W. C., Hunt, J. D., "The Use of Natural Anhydrite in Portland Cement," *ASTM Bulletin No. 161,* pg. 50–58 (1949).

[10] ASTM C157, "Standard Test Method for Length Change of Hardened Hydraulic-Cement Mortar and Concrete," *Annual Book of ASTM Standards,* Vol. 04.02, pp. 97–101 (1988).

[11] Kalousek, G. L., Jumper, C. H., Tregoning, J. J., "Composition and Physical Properties of Aqueous Extracts from Portland Cement Clinker Pastes Containing Added Materials," *Journal of Research, National Bureau of Standards,* Vol. 30, pp. 215–225 (1943).

[12] Odler, I., Becker, T., "The Effect of Some Liquifying Agents on Properties and Hydration of Portland Cement and Tricalcium Silicate Pastes," *Cement and Concrete Research,* Vol. 10, pp. 321–331 (1980).

[13] Sakai, E., Raina, K., Asaga, K., Goto, S., Kondo, R., "Influence of Sodium Aromatic Sulfonates on the Hydration of Tricalcium Aluminate With or Without Gypsum," *Cement and Concrete Research,* Vol. 10, pp. 311–319 (1980).

[14] Ramachandran, V. S., "Adsorption and Hydration Behavior of Tricalcium Aluminate-Water and Tricalcium Aluminate-Gypsum-Water Systems in the Presence of Superplasticizers," *Journal of American Concrete Institute,* pp. 235–241 (1983).

[15] Bruere, G. M., "Importance of Mixing Sequence When Using Set-Retarding Agents with Portland Cement," *Nature,* Vol. 199, pp. 32 (1963).

[16] Dodson, V. H., Farkas, E., "Delayed Addition of Set Retarding Admixtures to Portland Cement Concrete," *Proceedings, American Society for Testing and Materials,* Vol. 64, pp. 816–826 (1965).

[17] Dodson, V. H., "History of Darex Admixtures," *Construction Products Div., W. R. Grace & Co.,* In-house Publication, pp. 2–3 (1986).

[18] Burnett, I., "High Strength Concrete in Melbourne, Australia," *Concrete International,* Vol. 11, No. 4, pp. 17–25 (1989).

[19] ACI Committee 439, "Uses and Limitations of High Strength Steel Reinforcement," *American Concrete Institute, R-73,* (1973).

[20] Smith, G. L., Rad, F. N., "Economic Advantages of High-Strength Concretes in Columns," *Concrete International,* Vol. 11, No. 4, pp. 37–43 (1989).

[21] Madderom, F. M., "Excess Water Can Be a Costly Ingredient in Concrete," *Concrete Construction,* pg. 340 (1980).

[22] Basile, F., Biagini, S., Ferrari, G., Collepardi, M., "Effect of the Gypsum

State in Industrial Cements on the Action of Superplasticizers," *Cement and Concrete Research,* Vol. 17, No. 5, pp. 715–722, Sept. (1987).

[23] ASTM C143, "Standard Test Method for Slump of Portland Cement Concrete," *Annual Book of ASTM Standards,* Vol. 04.02, pp. 85–86 (1988).

[24] Ravina, D., Mor, A., "Consistency of Concrete Mixes-Effects of Superplasticizers," *Concrete International,* pp. 53–55 July (1986).

Chapter 4

SET ACCELERATING ADMIXTURES

INTRODUCTION

Set accelerating admixtures are best defined as those which, when added to concrete, mortar or paste, increase the rate of hydration of hydraulic cement, shorten the time of setting and increase the rate of early strength development. This class of admixtures is designated as Type C or Type E, and their performance requirements in portland cement concrete are specified in ASTM C494 [1]. Some of those requirements have been previously listed in Table 2-1.

SET ACCELERATING WATER REDUCING ADMIXTURES

The author disposes of the Type E, water reducing and accelerating admixture, quickly because there is no such product on the market today, in spite of what the various admixture manufacturers claim. Those few admixtures that are said to fall into the Type E category produce water reduction through the entrainment of unstable air and are simply mixtures of a set accelerator and an air entraining admixture (the latter will be discussed in Chapter 6).

Set Accelerating Admixtures

All of the effective components of the Type C admixtures are inorganic salts, with the exception of triethanolamine, which is an organic compound. The most often used of the inorganic salts is calcium chloride, $CaCl_2$; and most of the theories concerning the phenomenon of set acceleration caused by the inorganic salts have been based essentially on what is known about the effect of $CaCl_2$ on the rate of hydration of portland cement. Although we know what $CaCl_2$ (and possibly the other inorganic salts) does with respect to the performance of concrete, mortar and paste, a great deal of uncertainty still exists as to the mechanism that is involved. Because of its wide use, it has been the subject of many studies over the past 100 years. This particular salt is the oldest chemical admixture to be used in portland cement concrete.

Set Accleration By Calcium Chloride-Mechanism

As early as 1886, it was proposed that a calcium oxychloride, which was formed through the combination of calcium hydroxide, produced by the hydrating silicates, and the added $CaCl_2$, was responsible for the acceleration of time of setting [2].

$$Ca(OH)_2 + CaCl_2 + 11H_2O \rightarrow CaO \cdot CaCl_2 \cdot 12H_2O \quad (4\text{-}1)$$

A similar mechanism was proposed almost 50 years later, but the nature of the product of the reaction was different [3].

$$4Ca(OH)_2 + CaCl_2 + H_2O \rightarrow 4CaO \cdot CaCl_2 \cdot 14H_2O \quad (4\text{-}2)$$

However, both concepts were discarded when the products of the reactions pictured in Equations 4-1 and 4-2 were synthesized, and their respective X-ray diffraction spectra were compared with those of tricalcium silicate pastes, which had been hydrated in the presence of $CaCl_2$, and found to be absent [4].

Although it had been suggested that a reaction between $CaCl_2$ and the tricalcium aluminate phase to form a calcium chloroaluminate (Equation 4-3) might be responsible for the action of calcium chloride, the results of a kinetic study of the reaction between pure C_3A and $CaCl_2$, *in the presence of gypsum*, indicated that the rate of formation of the chloroaluminate was too slow to account for the reduction in time of setting and rapid development of early strength [5].

$$2C_3A + CaCl_2 + H_2O \rightarrow 2C_3A \cdot CaCl_2 \cdot H_2O \quad (4\text{-}3)$$

However, in the *absence* of gypsum, that same mixture; i.e., C_3A, water, and $CaCl_2$, exhibited a rapid uptake of the chloride ion from the aqueous phase and the composition of the product was calculated to be $2C_3A \cdot CaCl_2 \cdot 10H_2O$. The rate of chloride ion disappearance from the aqueous phase is shown in Figure 4-1. Although the hypothetical $C_3A \cdot CaCl_2 \cdot XH_2O$ compound had been the favorite of text books, the results of the work just cited support an earlier conclusion [6].

Studies of the composition of the portland cement-water aqueous phase have shown that there is very little, if any, decrease in its chloride ion content during the early stages of its hydration when the cement is properly retarded with calcium sulfate [7][8]. This is not to say that the chloroaluminate cannot or does not form when calcium chloride is present. For example, in 1979 the author ground a type I portland cement clinker with a sufficient amount of $CaCl_2$ to form the product described in Equation 4-3 (where $X = 10$). The resulting product exhibited the same time of setting and early strength characteristics as those of the same clinker that had been ground with sufficient gypsum to give normal setting and early strength properties. He concluded that if the cement contained less than the required amount of calcium sulfate, some of the C_3A could and/or would combine with $CaCl_2$, if it were present, and that the combination would serve to retard the reaction of the C_3A directly with water. He also reasoned that the amount of $CaCl_2$ tied up by the C_3A represented that quantity not able to cause acceleration of set and early strength development, which would explain why different portland cements respond differently to a given addition rate of $CaCl_2$.

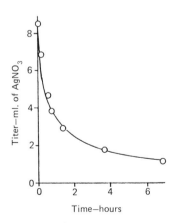

Figure 4-1. Rate of reaction of calcium chloride with tricalcium aluminate [5].

Another mechanism that had been proposed to explain the action of $CaCl_2$ had to do with the possible incorporation of the chloride ion in the tobermorite structure and thus activating the remaining anhydrated C_3S. This was abandoned when it was found that $CaCl_2$ did not react chemically with tricalcium silicate [9]. The results of some of the author's work in this area are summarized in Table 4-1. The chloride ion content of the aqueous phase of the C_3S-water-$CaCl_2$ mixtures held very closely to the theoretical value of 17.0 g/L, assuming no interaction between the $CaCl_2$ and C_3S. At the five hour period, the mixtures were thick and difficult to filter. The rise in chloride ion content of the aqueous phase is attributed to the water uptake by the hydrating C_3S and an increase in its concentration in the remaining unreacted water. The activation energies involved in the hydration of synthetic C_3S, with and without $CaCl_2$, have been compared and the results indicate that $CaCl_2$ does not change the activation energy of the hydration process and does not alter the mechanism of its hydration but only increases its rate of hydration [3].

Hydration of C_3S in Presence of Calcium Chloride

The influence of $CaCl_2$ on the hydration of synthetic C_3S is illustrated in Figure 4-2, wherein the isothermal conduction calorimeter curve of pure C_3S, in the presence of $CaCl_2$, is superimposed over that of the plain C_3S. (An interpretation of the calorimeter curve for plain C_3S was discussed in Chapter 1.) The addition of the $CaCl_2$ does not alter the initial heat evolution, but it does reduce the extent of the dormant period. If one can equate the area under the two temperature-

Table 4-1. Chloride Ion Uptake by Synthetic C_3S and Water.

MIXING PERIOD-MIN[a]	CHLORIDE ION IN AQUEOUS PHASE—GRAMS/LITER[b]
0.5	16.6
3	17.0
10	17.1
30	16.9
60	17.0
180	17.0
240	16.8
300	22.7

[a]$W/C_3S = 0.75$
[b]2% anhydrous $CaCl_2$ on weight of C_3S, theoretical chloride ion content = 17.0 g/L.

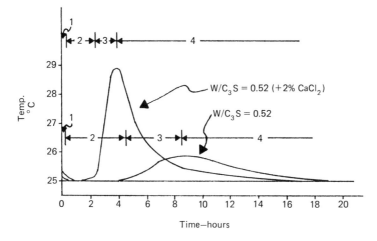

Figure 4-2. Isothermal conduction curves of pure C_3S/water and pure C_3S/water/ $CaCl_2$ mixtures.

time curves to the amount of C_3S hydrated, the amount of C_3S hydrated during the first 8-1/2 hours is increased by a factor of about 3 when calcium chloride is present.

A comparison of the rates of disappearance of the alite phase (C_3S) in a type 1 portland cement with and without calcium chloride is shown in Figure 4-3 [10]. Again, the $CaCl_2$ is seen to increase the amount

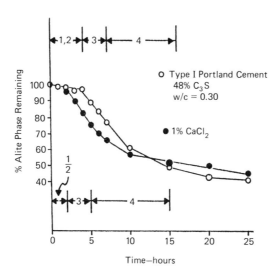

Figure 4-3. Effect of $CaCl_2$ on the amount of unreacted alite in Portland cement.

of alite hydrated over the first few hours. The following is a summary of the evidence thus far presented.

1. $CaCl_2$ does not react with the C_3S phase of portland cement and only causes it to hydrate more rapidly. This results in an acceleration of the time of setting, a larger than usual amount of C_3S hydration products and, consequently, a higher than normal early strength.
2. The C_3A phase of portland cement does not react with $CaCl_2$, *unless* the cement does not contain a sufficient amount of calcium sulfate to satisfy the needs of the aluminate phases. (The author referred to this type of cement in Chapter 1 as being "sulfate starved.")
3. The basic question dealing with the mechanism through which $CaCl_2$ activates the C_3S phase of portland cement has yet to be resolved.

In pursuing item 3 above, first consider the effect of $CaCl_2$ on the alkalinity, or pH, of the portland cement-water system. The author has found, as have others, that the presence of $CaCl_2$ in a portland cement-water mixture lowers the pH of the system [7]. Some of the data, illustrating this point is shown in Figure 4-4. In an effort to explain the data in Figure 4-4, the author measured the pH of the

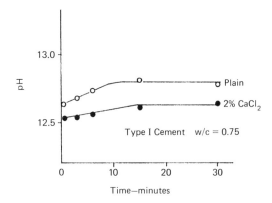

Figure 4-4. pH vs. time of a mixture of a Type I Portland Cement and water, with and without $CaCl_2$.

aqueous phase of mixtures of laboratory prepared C_3S and water, with and without $CaCl_2$ present. The results are shown in Figure 4-5. In addition to producing a lower pH in the aqueous phase, the presence of the $CaCl_2$ caused the pH of the system to drop much sooner; and the sudden decrease in pH of both mixtures corresponded favorably with the end of their dormant periods. The pH levels reported in Figure 4-5 for the synthetic C_3S are somewhat lower than those in Figure 4-4 for a type I portland cement because of the influence of the soluble alkali in the latter. The drop in pH can be attributed to the sudden formation of crystalline $Ca(OH)_2$ from a supersaturated condition to produce an aqueous phase saturated with calcium hydroxide. Thereafter crystallization proceeds rapidly [11].

The influence of calcium chloride on the $Ca(OH)_2$ content of synthetic C_3S-water aqueous phase is illustrated in Figure 4-6. The calcium hydroxide saturation level in the C_3S-water mixture was assumed to be that of $Ca(OH)_2$ in plain water, while that of the $CaCl_2$ mixture was established from a $Ca(OH)_2/CaCl_2$ combination in which the amount of the latter was adjusted to simulate the addition rate of the $CaCl_2$ to the C_3S-water mixture. The drop in the $Ca(OH)_2$ content of the aqueous phase occurs after 4 hours of mixing when $CaCl_2$ is present and after 8 hours when it is absent. Although it is generally agreed that more $Ca(OH)_2$ is produced in *pastes* during the early hydration of portland cement (and of synthetic C_3S) when $CaCl_2$ is present, it should be kept in mind that the data in Figure 4-6 only deals with the amount of $Ca(OH)_2$ soluble in the aqueous phase.

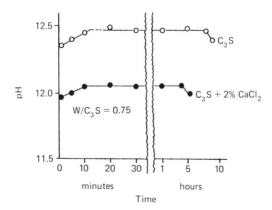

Figure 4-5. pH of C_3S-water mixtures with and without calcium chloride.

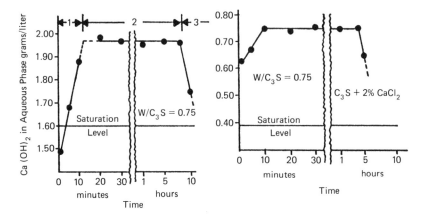

Figure 4-6. Influence of calcium chloride on the amount of calcium hydroxide in the aqueous phase of C_3S-water mixtures.

The author is inclined to agree with the proposal that:

- Because $CaCl_2$ makes the environment less alkaline, the C_3S-water was well as that of portland cement-water systems attempt to compensate for the reduced pH through an increase in rate of hydration. (There is probably some law of thermodynamics that applies here.)
- The reduced alkalinity affects the crystallization capability of the silicate hydration products leading to formation of very small crystals.
- The small crystals of the C_3S hydration product is responsible for the higher early strengths [12].

The influence of calcium chloride on the specific surface (or fineness) of the hydration products of synthetic C_3S has been shown to be very pronounced [3]. For example, the products of hydration in mixtures of pure C_3S and water, with and without $CaCl_2$ (2% on weight of C_3S) after an equal degree of hydration were found to have specific surfaces of 15.22 m^2/g for the water-C_3S system and 78.92 m^2/g when the system contained $CaCl_2$. The percent of free lime (CaO) and ignition loss were used as the criteria to determine equal degrees of hydration. The values cited for the two systems were obtained after 72 hours of hydration for the former and only after 8 hours of hydration for the latter.

Any factor that causes acceleration of time of setting, whether it be an admixture, high temperature, or high fineness of the portland cement, influences the particle size of the C_3S hydration product [13]. Under acceleration, very few of the C_3S hydration products can grow to long fibers and instead, the pore spaces are filled with many short fibrous crystals. Under normal hydration conditions, calcium silicate hydrates of a fibrous nature, 5–10 μm in length, can be detected in 7-day old pastes. When $CaCl_2$ is present, only a few calcium silicate hydrate fibers exceeding 1 μm in length can be found. It has also been reported that the compressive strength of pastes containing autoclaved calcium silicate hydrates, consisting of fine crystals with intergrowth, were 3 to 5 times stronger than those containing coarse crystals with very little intergrowth [14]. The author agrees that this phenomenon could well explain the high early strength development in concrete (as well as acceleration of time of setting) caused by $CaCl_2$, but it does not necessarily account for the later strengths of the concrete. This is discussed later in the chapter and also in Chapter 5.

Set Accelerating Salts Other Than Calcium Chloride

All of the experimental evidence indicates that calcium chloride acts as a catalyst (i.e., anything that alters the rate of a chemical reaction but does not take part in that reaction) in the hydration of pure C_3S and/or the alite phase of portland cement, causing the hydration products to be more numerous and smaller in size. Calcium chloride is not unique in its action in that there are a number of other calcium salts which exert a similar effect on the hydration of tricalcium silicate, but it is unique in that it is the most effective of the calcium salts, on a pound for pound basis.

Isothermal conduction calorimeter curves illustrating the effects of four different calcium salts on the rate of hydration of synthetic C_3S are illustrated in Figure 4-7. Each of the calcium salts was added to the C_3S, predissolved in the mix water, at an addition rate of 2% anhydrous salt, on the weight of C_3S. The lengths of the dormant periods, the times required to reach the maximum temperature of the second hydration peak and the temperature reached at the second peak of the hydration reaction are decidedly different for the four salts. The slopes of the temperature-time curves from the end of the dormant period to the peak temperature, relative to that of the untreated

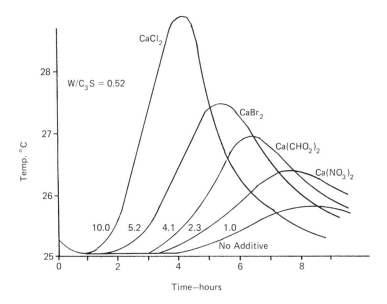

Figure 4-7. Isothermal conduction calorimeter curves for C_3S treated with various calcium salts.

C_3S, which was arbitrarily assigned a value of 1.0, are also indicated in Figure 4-7. With some stretch of the imagination, the two variables seem to be inversely related, almost linearly.

Using an isothermal conduction calorimeter somewhat different in design than that employed to generate the data in Figure 4-7 and at a lower water C_3S ratio, the relative effectiveness of several different calcium salts have been reported to be in the following decreasing order:

$$\text{bromide} > \text{chloride} > \text{thiocyanate} > \text{iodide}$$
$$> \text{nitrate} > \text{perchlorate} \quad [15]$$

Each of the salts was used at addition rates equivalent to 2% $CaCl_2 \cdot 2H_2O$ on the weight of C_3S. Both sets of data, the author's and that cited in the literature, clearly point to the fact that different calcium salts exert different effects on the rate of hydration of tricalcium silicate. While the author's work has been essentially confined to the study of calcium salts, the influence of other metal chlorides has been investigated, and these showed different degrees of influence on the dormant period of C_3S. In general, the effect of the cation was found to decrease in the following order:

$$Ca^{++} > Sr^{++} > Ba^{++} > Li^+ > Na^+ > K^+ \quad [15]$$

Either the cation is playing a part in the catalysis or providing some degree of set retardation through its influence on solubility or possibly ionic strength to modify the effect of the chloride ion.

The pH levels attained in the aqueous phase after 20 minutes of mixing synthetic C_3S-water combinations containing the four calcium salt depicted in Figure 4-7 (at the same addition rates and water-C_3S ratio) are listed in Table 4-2. All four calcium salts lowered the pH of the aqueous phase and in the order of their degree of acceleration of the C_3S hydration (Figure 4-7).

In 1970 the author examined various calcium salts for their relative accelerating effects on the time of setting of a type I portland cement-water paste [16]. Each of the salts was added as a part of the mix water at a rate of 2.0% (anhydrous) on weight of portland cement. The reduction in time of setting is plotted in Figure 4-8 as a function of the number of gram moles of calcium ion introduced. In general, the data points fall close to a straight line. With another cement, the data points could very well fall closer to or further from the arbitrarily drawn straight line. A similar relationship is noted when the number of gram moles of the anion is used as one parameter because the number of anion moles is twice that of the cation. One other facet of the data scatter in Figure 4-8 that must be considered is the precision of the method chosen for measuring the time of setting of the pastes.

Table 4-2. pH of the Aqueous Phase of C_3S-Water Mixtures Containing Various Calcium Salts.

ADMIXTURE[a]	pH OF AQUEOUS PHASE[b]
None	12.47
$Ca(NO_3)_2$	12.30
$Ca(CHO_2)_2$	12.25
$CaBr_2$	12.23
$CaCl_2$	12.05

[a]Each salt added—2% (anhydrous) on weight of C_3S.
[b]W-C_3S ratio = 0.75.

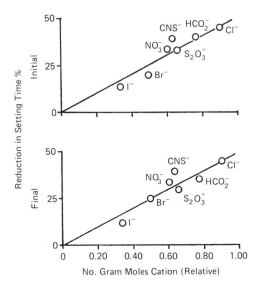

Figure 4-8.   Effect of various calcium salts on the time of setting of Portland Cement pastes.

INFLUENCE OF CALCIUM CHLORIDE ON PROPERTIES OF CONCRETE

Although calcium chloride is the most popular and most widely used of the inorganic set accelerating admixtures, it does produce a number of adverse effects on the properties of concrete that warrant mentioning. The first and foremost of these has to do with its potential corrosive action on metals embedded in the paste, mortar, or concrete. A great debate has been raging within the concrete (and admixture) industry over the past 10 years. One side questions whether or not the presence of the chloride ion introduced by the admixture is responsible for metal corrosion or whether the corrosion is due to the chloride ion introduced by the external application of chloride ion containing deicing salts. This had led to a second argument which is centered on what the maximum chloride level should be, whether introduced by the chloride containing admixture and/or by the deicing salts.

As a result of serious problems of corrosion of embedded steel in concrete associated with the use of $CaCl_2$ as an admixture, in 1972 the British Standards Institute (B.S.I.) of the United Kingdom, imposed the restriction that no more than 1.5% anhydrous $CaCl_2$, on the weight of cement, be used in concrete containing embedded metal and that it was not to be used with certain cements, such as high alumina, sulfate resisting or those classed as super-sulfated. In 1977, the B.S.I. went one step further and recommended that $CaCl_2$ *not* be added to concrete containing embedded metal. At, or about, this time the American Concrete Institute began to consider establishing limits on the chloride ion content of concrete, depending upon its use and environment.

After studying all of the available data, the American Concrete Institute (Committee 201 and 212) established the limits, as shown in Table 4-3, on the "water soluble" chloride ion content of concrete prior to service exposure [17]. The adjective "water soluble" preceding chloride requires some clarification. For example, the author has found that when a concrete made from a type I portland cement and treated with 2.0% $CaCl_2$ (anhydrous) and moist cured for 28 days, was ground to four different fineness values (including the aggregate), different amounts of soluble chloride ion were found after repeated water extractions. The analytical data are shown in Table 4-4. The author showed earlier in this chapter that the chloride ion does not act by combining with either the C_3S or the C_3A (if the cement is properly sulfated) phases of portland cement. That amount of chloride ion that remains uncombined is considered to be the soluble chloride ion in the concrete. Conflicts in experimental results seem to be the name of the game, and the concept of soluble chloride ion is no exception. For example, it has been reported that as much as 75% to 90% of the chloride added, via the admixture, ends up as chloroal-

Table 4-3.   ACI Recommendations for Chloride Ion Content of Concrete (ACI 212.1R-81) [17].

CATEGORY OF CONCRETE SERVICE	MAX. WATER SOLUBLE CHLORIDE ION[a]
Prestressed Concrete	0.06
Conventionally reinforced concrete in a moist environment exposed to chloride	1.10
Conventionally reinforced concrete in a moist environment but not exposed to chloride ion	0.15
Above ground construction where the concrete will stay dry	No limit

[a]% on weight of cement.

Table 4-4. Amount of Soluble Chloride Ion in Concrete.

FINENESS OF CONCRETE—cm^2/g^a	AMOUNT OF SOLUBLE CHLORIDE FOUND—% ON WEIGHT OF CEMENT[b]
3050	0.88
4160	0.97
4910	1.06
5550	1.18

[a]Concrete contained 517 lbs/yd^3 of cement with 2% anhydrous $CaCl_2$ added on weight of cement.
[b]Theoretical amount of Cl^- ion present = 1.28% on weight of cement.

uminate and that the chloride ion becomes combined even in a cement that contains no C_3A [18].

The amount of chloride ion that exists in the soluble category depends on (1) its original addition rate, (2) composition of the cement, (3) the age of the concrete, and (4) the fineness to which the concrete is ground prior to extraction with water (Table 4-4).

Those experts in the U.S. who argue for a limitation of soluble chloride added by way of the chemical admixture have made a number of claims, only a few of which will be cited here [19].

1. The repair of corroded structures can cost 4 to 5 times as much as the original structure.
2. If the chloride ions from the environment intrude the concrete, any chloride that has been intentionally added will augment or further promote the corrosion of the metal embedded in the concrete.
3. Strict limits are needed as a safeguard against corrosion where the concrete is inadequately compacted, allowing the embedded metal to be exposed to the intrusion of oxygen and moisture, both of which are needed for corrosion, with or without the presence of the chloride ion.
4. Where the concrete is subject to sulfate ion attack and/or alkali-aggregate reaction, the presence of the chloride ion aggravates the deleterious effects of those reactions on the integrity of the concrete.

Those opposed to the chloride ion limitation offer a number of rebuttals [19]. Only a few of these are listed as follows:

1. The substitution of non-chloride set accelerators for calcium chloride will cost $2 to $6 more per cubic yard of concrete.

2. If the proper concrete placing and curing practices are strictly enforced, the penetration of the chloride ion from deicing salts will be minimal. Metal reinforced concrete containing up to 2% calcium chloride, on weight of cement, has been proven to be trouble free, where adequate cover of the metal is provided.
3. The amounts of soluble chloride ion, which contributes to the corrosion that is introduced by chloride containing chemical admixtures are very small compared with that of potential environmental sources.
4. Calcium chloride is very effective, because it reduces the time of setting, permits earlier finishing, reduces the curing time and produces higher earlier strength with less energy consumption than steam curing. It also extends the construction season and makes earlier use of the structure possible.
5. Corrosion of embedded metal requires three elements: water, oxygen, and metal. While chloride ions may accelerate the process, they do not cause it.

The author's personal and professional opinion with regard to the five rebuttals just listed are outlined in Table 4-5.

Some of the other negative effects resulting from the use of $CaCl_2$ in concrete include:

- Reduced compressive strengths at later ages (vs. those of the plain reference concrete). This is why the minimum compressive strength of concrete treated with a type C admixture is set at 90% of the reference at 6 months and 1 year [1].
- High early shrinkage, but at 90 days the shrinkage of the treated and untreated concretes is essentially the same.
- Reduced response to air entrainment (Chapter 6).
- Reduced sulfate resistance (Chapter 6).
- Blotching of the concrete surface, caused by an interaction of the chloride ion with the alkali and C_4AF in the cement when both of the latter are high.

NON-CHLORIDE, NON-CORROSIVE SET ACCELERATING SALTS

The author began research in 1962, aimed at developing a non-chloride, non-corrosive concrete accelerating admixture and calcium formate ($Ca(CHO_2)_2$) was found to meet all the requirements of a type

Table 4-5. Author's Opinion of the Rebuttals to the Use of Calcium Chloride Containing Set Accelerating Admixtures.

REBUTTAL NO.	AUTHOR'S OPINION
1	Not necessarily so, but even if this is true, one has to put a price on the integrity of the concrete.
2	This is so if the adequate cover of the metal means that it is coated with a material such as an epoxy resin. However, the author has seen too many of such treated rebars with cracks in their coating, due to careless handling in the field, and these points of metal exposure are susceptible to corrosion. If the term, adequate coverage, refers to proper placement and compaction, forget it.
3	This may be so, but only in the case of water reducing admixtures that contain $CaCl_2$ (Table 3-1). Even then the amount of potential soluble chloride ion can exceed that set forth in Table 4-3.
4	The author agrees, but the same claims can be made for the use of non-chloride containing set accelerating admixtures.
5	This is essentially true because the added electrolytic action of the chloride ion in the concrete enhances the tendency for corrosion. But, in today's concrete industry how can one be assured that the concrete surrounding the embedded metal is properly placed and compacted?

C admixture, and did not corrode embedded metal in concrete treated with the admixture [20]. Although calcium formate is not as effective as an accelerator of time of setting as calcium chloride, on a pound for pound basis, it does produce the satisfactory increases in early strength as well as that at later ages. Typical setting time data for concrete, with and without calcium formate, are shown in Figure 4-9 [21]. The average compressive strengths of concretes made from six different type I portland cements are listed in Table 4-6, expressed as percentage of the plain concretes, i.e., reference compressive strength is 100.

The answer to the question concerning the mechanism through which calcium formate functions as a set accelerator is still undecided (as it is for calcium chloride). Although the formate is not as good an accelerator of set as the chloride, its effect on the later strength of the treated concrete is less depressing than that of $CaCl_2$. This has been observed by others [23]. Perhaps the mechanisms are the same, and the higher later compressive strengths produced by the formate

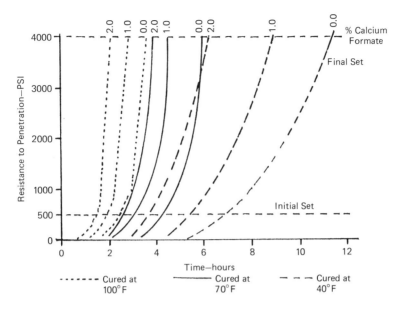

Figure 4-9. Influence of calcium formate on the time of setting of concretes cured at three different temperatures.

are the results of less set acceleration (this will be discussed later in this chapter). One other observation warrants recognition. Calcium silicate hydrate formation is greatest in portland cement pastes containing $CaCl_2$, and ettringite (and/or ettringite-like) compound formation is greatest in pastes containing the formate especially, when the cement is undersulfated [24]. Because the setting of concrete is primarily influenced by the degree of hydration of the C_3S phase, perhaps the consumption of the formate ion by the C_3A phase leaves less of it available to activate the C_3S.

Table 4-6. Influence of Calcium Formate on the Compressive Strength of Concrete.[a]

| | COMPRESSIVE STRENGTH—% OF REFERENCE | | | |
CALCIUM FORMATE[b]	1-DAY	7-DAYS	28-DAYS	1-YEAR
1.0	136	123	115	110
2.0	152	131	107	103

[a]C.F. = 564 lb, W/C = 0.57, prepared and cured in accordance with ASTM C192 [22].
[b]% of solids on weight of cement.

The biggest problem in using calcium formate as a set accelerating admixture is its low solubility in water. At room temperature, its maximum concentration in water is only about 14% by weight. Therefore, it must be added to concrete in the solid form which creates dispensing problems. Its companion compound, sodium formate, which is also a good set accelerator, is approximately 3 times more soluble in water, but the introduction of the added alkali can cause problems (Chapter 6).

A second non-chloride, non-corrosive, set accelerating admixture based on calcium nitrite $(Ca(NO_2)_2)$ was introduced in 1968 and patented in 1969 [25]. Its acceptance by the concrete industry, at the time of its introduction, was poor because of its cost (its only source was in Japan), although it had been reliably established that it was non-corrosive to embedded metals and that it met all the requirements of a type C admixture [1]. It was not until 1979, after five years of intensive research, that it was determined that calcium nitrite was also a very effective corrosion inhibitor for metals embedded in concrete. Since 1979, calcium nitrite has found considerable acceptance for use in steel-reinforced, post-tensioned and prestress concrete, which is apt to come in contact with chlorides from deicing salts or marine environment. In addition, its cost dropped appreciably due to the creation of a manufacturing facility in the U.S. in 1983.

Why is corrosion inhibition so important? Just look at the statistics. As of this writing, of the more than 560,000 highway and concrete bridges in the U.S., nearly one-half have been branded as being structurally unsound or functionally obsolete. In most cases, corrosion of the reinforcing metal is the principal cause for loss in structural integrity. It has been estimated that replacing and/or repairing these structures will cost the U.S. taxpayer 6 to 10 billion dollars and that dollar value continues to escalate with time. Concrete bridge decks which are, as a rule, reinforced with iron or its alloys, are subject to large amounts of chloride containing deicing salts during the winter months and are especially susceptible to deterioration by the corrosion of the embedded metal. When the metal reinforcement is based on iron, it becomes oxidized to ferric oxide (or one of its hydrates) as shown in Equation 4-4.

$$4Fe + 3O_2 \quad \overset{CaCl_2}{\underset{H_2O}{\rightarrow}} \quad 2Fe_2O_3 \qquad (4\text{-}4)$$

The increase in volume due to the formation of ferric oxide (rust) is almost twice that of the iron, and this results in expansive disruption of the concrete surrounding it. This leads to cracking, spalling, and discoloration of the concrete by the reddish brown ferric oxide. The cracking and spalling permits more oxygen, moisture, and chloride intrusion, and the whole process takes place at an ever increasing rate, almost logarithmic with time.

The reader should not get the impression that calcium nitrite is the panacea in corrosion protection, but both laboratory and field data indicate that it will provide more than an order of magnitude of reduction of corrosion rate in severe salt environments. If, for example, bridge decks and/or concrete floors in a parking garage, both of which are subject to exposure to chloride containing deicing salts, normally show the effects of corrosion in 5 to 10 years, an integrity life span of 30 to 50 years (or even more) can be expected when the concrete contains calcium nitrite.

Various viewpoints cited earlier were expressed by both sides of the great chloride debate. Here is one more piece of kindling wood for the fire. In a recent survey, 137 architectural and engineering firms throughout the U.S. were asked, among other things, whether they prohibited the use of chloride-based admixtures in their concrete mix designs [26]. Ninety percent said that they had banned the use of such admixtures in all concrete except certain foundation portions.

Because of the accelerating effect of time of setting of concrete produced by calcium nitrite, the addition of a set retarding admixture (to be discussed in Chapter 5) is almost mandatory in warm weather in order that the concrete can be delivered and properly placed. Since the double addition of admixtures often places a burden on the concrete producer, a neutral set version of calcium nitrite was developed in 1984 [27].

Following the 1965 patenting of calcium formate as a non-chloride, non-corrosive set accelerator, a number of other inorganic/organic salts were suggested as similar chemical admixtures. Spodumene, $LiAlSiO_6$, a naturally occurring mineral, was found to be an excellent accelerator of set and an enhancer of early compressive strength [28]. The same investigators found that the use of lithium oxalate, $Li_2C_2O_4$, an alkali metal salt of oxalic acid (organic), would perform in essentially the same way [29]. The use of spodumene and lithium oxalate was never put into practice because of the uncertainties surrounding the effect of the lithium ion, which is an alkali metal ion, on possible alkali-aggregate reactions (see Chapter 7). Calcium nitrate, $(Ca(NO_3)_2)$, was proposed as a basic component of a set accelerating admixture, in conjunction with triethanolamine, in 1981 [30]. As was the case with the lithium salts, no claim was made by the developers that the admixtures would inhibit corrosion.

The most recent non-chloride, non-corrosive (but not corrosion inhibiting) set accelerating admixture to be introduced to the U.S. concrete market is the alkali, alkaline earth metal and ammonium salts of thiocyanic acid [31]. It is doubtful if either the alkali metal salt or the ammonium salt will ever be popular because of the adverse effects of the added alkali through the use of the former and because of the strong ammonia odor of the concrete treated with the latter. Like the nitrate, the thiocyanate must be accompanied by an alkanolamine in order to attain the desired end results. The combination of the thiocyanate salts with those of nitric acid and an alkanolamine was patented in 1984 [32].

ALKANOLAMINE SET ACCELERATORS

Alkanolamines are generally strong organic bases and probably the most popular of these is triethanolamine (TEA). It first found its way into concrete in 1934, when it was interground as an addition with portland cement in combination with calcium lignosulfonate (CLS). The purpose of the combination was to increase the early compressive strength of the concrete made from the cement [33]. The addition rate of the TEA, on weight of cement (0.001%) was so small that its full accelerating effect on the hydration of the portland cement was not realized. The necessary strength gains were attained but through water reduction (Chapter 3) caused by the CLS component.

In 1958 a combination of TEA and CLS was placed on the concrete market as a water reducing-strength enhancing admixture. This time the TEA was added at a rate of 0.02% on weight of cement, and its effect on time of setting and on early strength was very noticeable. Since then TEA has become a common ingredient of water reducing admixtures mainly to partially off-set the set retardation caused by the cement dispersing component of the admixture and is generally preferred to calcium chloride because of the potential corrosive effect of the chloride ion.

Unlike the various inorganic salts previously discussed, the response of portland cement to a variation in the addition rate of TEA is at best fickle. For example, as the addition rate of the inorganic salts is increased, the degree of set acceleration is increased, but not necessarily linearly, while that of TEA can produce either set acceleration or retardation. The most popular addition rate of TEA is in the range of 0.010% to 0.025% on the weight of cement. The mechanism through which TEA acts is not clearly understood because of its rather erratic influence on the time of setting and early strength of concrete. To the author's knowledge, TEA has never been used as a set accelerating admixture as such but always in combination with a dispersing agent (Chapter 3) to partially off-set the set retardation caused by the adsorbed dispersant.

The time of setting characteristics of concrete made from a type I portland cement using four different addition rates of TEA are shown in Figure 4-10 [21]. The data indicate that at an addition rate of 0.02%,

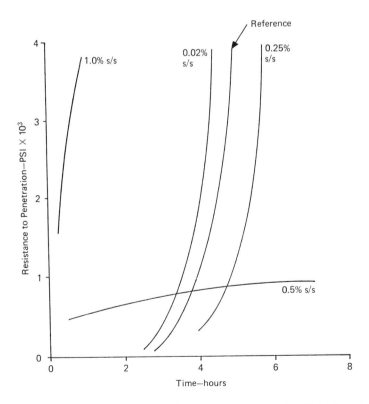

Figure 4-10. Time of setting characteristics of concretes treated with triethanolamine.

TEA acts as an accelerator of set, at 0.25% as a mild set retarder, at 0.5% a severe retarder and at 1.0% a very strong accelerator, producing a condition similar to flash set [34]. The 24-hour compressive strength of concretes treated with more than 0.03% TEA typically exhibit very little gain while the lower concentrations produce higher compressive strengths than those of the untreated concretes.

The hydration characteristics of synthetic C_3S have been studied in the presence of 0.1%, 0.5%, and 1.0% TEA, and the results indicate that at these addition rates, the TEA increases the length of the dormant period but enhances the surface area of the hydrated calcium silicate products [35]. The influence of the addition rate of TEA on the dormant period of the hydration of synthetic C_3S had been noticed earlier in the author's laboratory, but never reported, and some of these data are illustrated in Figure 4-11. When the addition rates of TEA of 0.5%, 1.0%, 5.0%, and 10% were used in a study of the response of tricalcium aluminate-water and tricalcium aluminate-gypsum-water systems, it was noted that the rate of hydration of the aluminate phase was increased as the amount of TEA was increased and that the formation of ettringite, when gypsum was present, was accelerated [36]. However, in both of the studies just cited, higher

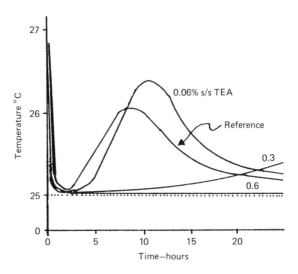

Figure 4-11. Isothermal conduction curves of synthetic C_3S in the presence of triethanolamine.

than normal TEA addition rates were employed. This leads the author to conclude that much of the confusion in today's literature is the result of differences between the addition rates chosen for study and those used in normal concrete practice.

TEA has the ability to chelate certain metallic ions in a highly alkaline medium; i.e., above a pH of 12, one of which is the ferric ion (Fe^{+3}). The average Fe_2O_3 content of typical type I, II, and III cements is 2.7%, most of which resides in the C_4AF phase of the cement. This equates to about 1.9% Fe^{+3} in the cements. It was postulated as early as 1958 that Fe^{+3} precipitates during the hydration of portland cement, coating the silicates and aluminates with gelatinous, water impermeable ferric hydroxide ($Fe(OH)_3$), thereby slowing down their hydration [37].

The results of chemical analyses of the aqueous phase of portland cement-water mixtures, with and without TEA, have shown that when TEA is present, more Al^{+3} and Fe^{+3} ions are present in solution [38]. This tells the author that in the presence of TEA, less of the insoluble $Fe(OH)_3$ forms which can cause set retardation by its deposition on the cement phases and that coupled with the added solubility of the Al^{+3} ion favor a more rapid formation of ettringite (or its low sulfate form). In the author's opinion, the TEAs chelating action on the Fe^{+3} ion makes the C_4AF phase act like C_3A and thus increases the rate of calcium sulfoaluminate formation. Since ettringite is a contributor to initial and final time of setting (and to some extent to early strength), TEA, at the proper addition rate, will counteract a part of the set retardation of the dispersant present in water reducing admixtures. The proposed mechanism for the chelation of the ferric ion (Fe^{+3}) by TEA is illustrated in Figure 4-12 [39].

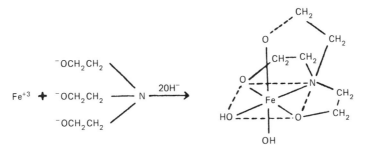

Figure 4-12. Chelation (solubilization) of the ferric ion in Portland Cement by triethanolamine.

The results of an experiment conducted by the author in the late 1970s are summarized in Table 4-7 and illustrated in Figures 4-13, 4-14, and 4-15. The data indicate that the degree of set acceleration through the addition of a type C admixture is almost linearly related to the increase in 3-day compressive strength and inversely related to the compressive strength of the treated concrete at 28 days. This concept of time of setting as a function of early and later compressive strengths will be further discussed in Chapter 5.

Table 4-7. Influence of Type C Accelerating Admixtures on the Early and Later Compressive Strengths of Concrete.[a]

ADMIXTURE	ADDT'N RATE[b]	DECREASE IN FINAL TIME OF SETTING-MIN	INCREASE IN COMPRESSIVE STRENGTH—%[c]	
			3-DAY	28-DAY
$CaCl_2$	1.0%	90	136	109
$CaCl_2$	2.0%	125	148	102
$Ca(NO_2)_2$	1.0%	65	128	113
$Ca(NO_2)_2$	2.0%	90	137	112
$Ca(HCO_2)_2$	1.0%	70	123	115
$Ca(HCO_2)_2$	2.0%	100	133	105

[a]C.F. = 517 lb, W/C = 0.55, type I portland cement.
[b]% solids of anhydrous salt, on weight of cement.
[c]Reference concrete = (100).

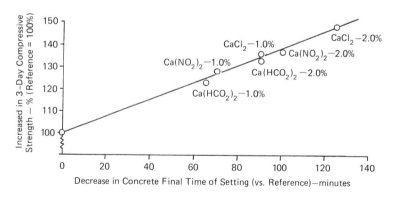

Figure 4-13. Relationship between acceleration of final time of setting and 3-day compressive strength of concrete treated with various concrete accelerating admixtures.

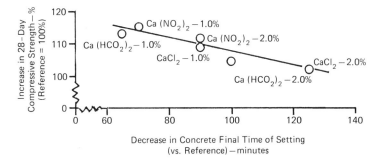

Figure 4-14. Relationship between acceleration of final time of setting and 28-day compressive strength of concrete treated with various concrete accelerating admixtures.

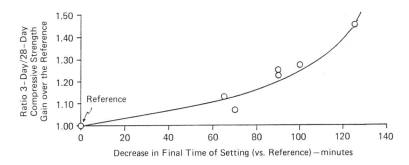

Figure 4-15. Relationship of final time of setting of concretes treated with various set accelerating admixtures and their ratios of 3-day to 28-day compressive strength gains over the reference (from Table 4-7).

ECONOMICS IN USING SET ACCELERATING ADMIXTURES

First consider the economics of using set accelerating chemical admixtures in cold weather, wherein their major use lies. Although the reactions between the various phases of portland cement and water (and calcium sulfate) are very complex, they do tend to obey the law of kinetics as expressed by Arrhenius; i.e., for every 18° F (10° C) rise in temperature, the rate of reaction increases by a factor of 2 or 3. Conversely, for every 18° F decrease in temperature, the rate of a chemical reaction is reduced by a similar factor. Therefore, in order to overcome the depression of the hydration rate of portland cement in cold weather, a set accelerating admixture is frequently added.

Set accelerating admixtures, especially calcium chloride, are referred to as anti-freeze additives for concrete. The author was taught that an anti-freeze agent was something that would dissolve in a fluid and lower its freezing point and that every gram mole of the solute would lower the freezing point of 1000 grams of the solvent by 3.35° F (1.85° C). Suppose one has a concrete whose C.F. is 500 lbs and whose W/C is 0.55 and 2% anhydrous calcium chloride is added to accelerate its time of setting. The total amount of $CaCl_2$ would be 10 lbs and represent 40.90 gram moles (10 lb. × 454 g/lb ÷ 111 (molecular weight)). The amount of water, or solvent, in the mixture is 124,850 g (500 lbs × 0.55 × 454 g/lb). This means that the molar concentration of the $CaCl_2$/1000 grams of water is 0.328. Assuming that all of the $CaCl_2$ dissolves in the concrete aqueous phase, the lowering of its freezing point would only be 1.09° F (3.35 × 0.328). Considering the results of the previous mathematical exercise, the admixture appears to offer very little anti-freeze protection to the concrete. One can argue that the early hydration of the cement will take up a part of the water which serves to concentrate the amount of $CaCl_2$ in the unreacted water located in the pores and capillaries of the concrete. If one assumes that 30% of the originally added water is consumed during the early hydration, the increase in $CaCl_2$ concentration will only lower the freezing point of the remaining water by 1.56° F.

The principal objective of using a set accelerator in cold weather is twofold: (1) to give the portland cement in the concrete an opportunity to hydrate to a point that if the remaining water in the pores freezes, it has sufficient volume of empty pores into which it can expand, and (2) the concrete has sufficient strength to resist damage by the expansion of the freezing water. These are basically the same goals set forth by the American Concrete Institute for the producers of concrete during cold weather [40].

The concrete producer and/or engineer has basically four options when it comes to cold weather concreting.

1. Switch from a type I to type III portland cement, which will cost the contractor money (on the average $0.02/lb of cement). It would also cost the concrete producer too, if he/she had no storage silo specifically assigned to a type III cement.
2. If the concrete producer cannot accommodate the type III cement, the amount of type I cement can be increased in the con-

crete by 100 to 150 lb/yd^3. This will cost the producer about $3.00/yd^3$, which is passed on to the contractor.

3. If the concrete producer chooses to use a type I cement at the usual C.F., he/she will have to bear the costs of using warm water and aggregates. This will add approximately $3.00/yd^3$ to the costs which are passed on to the contractor. In addition, the contractor will probably have to provide heating blankets for the placed concrete. This additional cost has been estimated to be $1.00/yd^3$ of concrete.

4. The contractor can elect to have the concrete delivered at a temperature of 35° F–40° F, which will reduce the cost of warming the raw materials and contain a chemical set accelerating admixture. This will probably cost the contractor $2.25/yd^3$.

No matter how one cuts the cake, it is less costly to use the normal amount of a type I cement along with a set accelerating admixture, even though a minimum protection is required against the cold blasts of winter.

The economics involved in the use of set accelerating admixtures at normal, or even high temperatures, is more difficult to orchestrate. Generally, when they are used, under these conditions, for the purpose of developing high compressive strength much sooner than normal, earlier stripping of concrete forms or release, in the case of prestressed concrete, is realized. All of this leads to an increase in the use-rate of the concrete forms, a reduction in the amount of heat applied and a more rapid erection of a multi-story construction. Again, the concrete producer and/or contractor have several options.

1. Increase the C.F. of the concrete which will cost $3.00/yd^3$ of concrete.

2. Changes from a type I to a type III cement, which will increase the price of the concrete by $3.00/yd^3$.

3. Apply an external source of heat, such as steam curing in the case of precast and/or prestressed concrete which can cost between $3.00 and $3.50/cubic yard.

4. Add a set accelerating admixture, the cost of which will vary between $2.00 to $2.25/yd^3$ of concrete.

5. If form removal and release time (in the case of prestressed concrete) are of no concern to the producer/contractor, he/she can

reduce the C.F. of the concrete and add a water reducing admixture in conjunction with the set accelerator. The former will off-set the negative effects of the latter on later strengths (Figure 4-14).

REFERENCES

[1] ASTM C494, "Standard Specification for Chemical Admixtures for Concrete," *Annual Book of ASTM Standards*, Vol. 04.02, pp. 245–252 (1988).

[2] Candelot, E.,: Zement mit Schneller Bindezeit," *Mónieture Industrial Belgium*, Vol. 13, pg. 128 (1886).

[3] Koyanagi, K., "Das Abbinden and Erharten das Portland-Zements," *Zement*, Vol. 23, pg. 705 (1934).

[4] Celani, A., Collepardi, M., Rio, A., "The Influence of Gypsum and Calcium Chloride on the Hydration of Tricalcium Silicate," *L'Indistrià Italiana del Cemento*, Vol. 36, pp. 669–678 (1966).

[5] Rosenberg, A. M.,"Study of the Mechanism Through Which Calcium Chloride Accelerates the Set of Portland Cement," *Journal of American Concrete Institute*, pp. 1261–1269 (1964).

[6] Wells, L. S., "Reaction of Water on Calcium Aluminates," *Journal of Research, National Bureau of Standards*, Vol. 1, pg. 951, Dec. (1928).

[7] Roberts, M. H., "Effect of Admixtures on the Composition of the Liquid Phase and the Early Hydration Reactions in Portland Cement Pastes," *British Royal Society, Paper No. 61*, pp. 210–219 (1968).

[8] Dodson, V. H., "Composition of Cement-Water Aqueous Phase in the Presence of Admixtures," *RILEM Presentation*, (1968).

[9] Richartz, W., "The Combining of Chloride in the Hardening of Cement," *Zement-Kalk-Gips*, No. 10, pg. 447 (1969).

[10] Angstadt, R. L., Hurley, F. R., W. R. Grace Research Center, *Private Communication* to V. H. Dodson (1969).

[11] Greene, K. T., "Early Hydration Reactions of Portland Cement," *Proceedings of the Fourth International Symposium on the Chemistry of Cement*, Washington, DC, pp. 359–374 (1960).

[12] Kurczyk, H. G., Schwiete, H. E., "Electron Microscopic and Thermochemical Investigations on the Hydration of Calcium Silicates $3CaO-SiO_2$ and β 2 $CaO-SiO_2$ and the Effects of Calcium chloride and Gypsum on the Process of Hydration," *Tonindustria-Zeitunq*, Vol. 84, pp. 585–598 (1960).

[13] Richartz, W., "Electron Microscope Investigations About the Relations Between Structure and Strength of Hardened Cement," *Fifth International Symposium on the Chemistry of Cements*, Japan, Paper No. III-125 (1968).

[14] Mindess, S., "Relation Between Compressive Strength and Porosity of Autoclaved Calcium Silicate Hydrates," *Journal of the American Ceramic Society*, Vol. 53, No. 11, pp. 621–624 (1970).

[15] Kantro D. L., "Tricalcium Silicate Hydration in the Presence of Various Salts," *Journal of Testing and Evaluation*, Vol. 3, No. 4, pp. 312–321 (1975).

[16] ASTM C266, "Standard Test Method for Time of Setting of Hydraulic-Cement Paste by Gillmore Needles," *Annual Book of ASTM Standards*, Vol. 04.01, pp. 242–245 (1986).

[17] ACI 212.1R81, "Admixtures for Concrete," *American Concrete Institute*, Detroil (1981).

[18] Verbeck, G. L., "Mechanisms of Corrosion of Steel in Concrete," *Corrosion of Metals in Concrete*, Publication SP-49, American Concrete Institute, Detroit, pp. 21–38 (1975).

[19] Anonymous, "Pros and Cons of Chloride Limits: A Comparative Summary," *Concrete Construction*, pp. 811–813, Oct. (1982).

[20] Dodson, V. H., Farkas, E., Rosenberg, A. M., "Non-Corrosive Accelerator for Setting of Cements," *U.S. Patent No. 3,210,207*, Oct 5 (1965).

[21] ASTM C403, "Standard Test Method for Time of Setting of Concrete Mixtures by Penetration Resistance," *Annual Book of ASTM Standards*, Vol. 04.02, pp. 210–213 (1988).

[22] ASTM C192, "Standard Practice for Making and Curing Concrete Test Specimens in the Laboratory," *Annual Book of ASTM Standards*, Vol. 04.02, pp. 110–116 (1988).

[23] Gebler, S., "Evaluation of Calcium Formate and Sodium Formate as Accelerating Admixtures for Portland Cement Concrete," *Journal of American Concrete Institute*, Sept/Oct., pp. 439–444 (1983).

[24] Bensted, J., "Early Hydration Behavior of Portland Cement in Water, Calcium Chloride and Calcium Formate Solutions," *Silicates Industries*, pp. 67–69 (1980).

[25] Angstadt, R. L., Hurley, F. R., "Accelerator for Portland Cement," *U.S. Patent No. 3,427,175*, Feb 11 (1969).

[26] Phelan, W. S., "Non-Chloride Accelerators: Some Case Histories," *Concrete Construction*, Oct., pp. 789–790 (1982).

[27] Dodson, V. H., Hayden T. D., "Corrosion Inhibiting Additive for Cement Compositions," *U.S. Patent No. 4,466,834*, Aug. 21 (1984).

[28] Angstadt, R. L., Hurley, F. R., "Spodumene Accelerated Portland Cement," *U.S. Patent No. 3,331,695*, July (1967).

[29] Angstadt, R. L., Hurley, F. R., "Hardening Accelerated Portland Cement," *U.S. Patent No. 3,373,048*, March (1968).

[30] Tokay, V., "Additive Composition for Portland Cement Materials," *U.S. Patent No. 4,337,094*, June (1982).

[31] Rosskopf, P. A., "Additive for Hydraulic Concrete Mixes," *U.S. Patent No. 4,373,956*, Feb. 15 (1983).

[32] Gerber, H. A., "Admixture for Hydraulic Cement," *U.S. Patent No. 4,473,405*, Sept. 25 (1984).

[33] Tucker, G. R., Kennedy, H. L., Renner, M. S., "Concrete and Hydraulic Cement," *U.S. Patent No. 2,031,621*, Feb. 25 (1936).

[34] Rosenberg, A. M., Construction Materials Division of Dewey & Almy, W. R. Grace, *Private Communication* to V. H. Dodson, December 4 (1968).

[35] Ramachandran, V. S., "Influence of Triethanolamine on the Hydration Characteristics of Tricalcium Silicate," *Journal of Applied Chemistry and Biotechnology*, Vol. 22, pp. 1125–1138 (1972).

[36] Ramachandran, V. S., "Action of Triethanolamine on the Hydration of Tri-

calcium Aluminate," *Cement and Concrete Research,* Vol. 3, No. 1, pp. 41–54 (1973).

[37] Steinour, H. H., "The Setting of Portland Cement," *Research and Development Laboratories, Portland Cement Association,* Bulletin 98, pp. 13–14 (1958).

[38] Kalousek, G. L., Jumper, C. H., Tregoning, J. J., "Composition and Physical Properties of Aqueous Extracts from Portland Cement Clinker Pastes Containing Added Materials," *Journal of Research, National Bureau of Standards,* Vol. 30, pp. 215–225 (1943).

[39] Chaberek, S., Martell, A. E., "Organic Sequestering Agents," John Wiley & Sons, Inc., New York, NY, pg. 325 (1959).

[40] ACI Committee 306, "Standard Specifications for Cold Weather Concreting," *ACI 306.1-87, American Concrete Institute,* Detroit (1987).

Chapter 5

SET RETARDING CHEMICAL ADMIXTURES

INTRODUCTION

A set retarding admixture is defined as one that delays the time of setting of portland cement paste and hence that of its mixtures, such as mortars and concrete [1]. Consequences of this delay in the rate of hardening, or setting, include a delay in the development of the early strength of the concrete, mortar, or paste and an increase in later compressive strength of the respective cementitious masses. There are three types of retarding admixtures recognized by ASTM: Type B, which simply retards the hydration of the portland cement; Type D, which not only retards the hydration but also acts to disperse the cement and thereby provide water reduction; and finally Type G, which is a high range water reducing and set retarding admixture [2]. Some of the physical properties of concrete, specified by ASTM, treated with the three types of chemical set retarding admixtures have been previously listed in Table 2-1 (p. 24).

ASTM TYPE B SET RETARDING ADMIXTURES

Very few, if any, of the Type B admixtures are on the market today, having been replaced by the more sophisticated bifunctional water reducing set retarders. At the height of their popularity in the 1930s, the most commonly used of these were based on water soluble salts such as sodium metaborate ($Na_2B_2O_4$) or tetraborate ($Na_2B_4O_7$), stannous sulfate ($SnSO_4$), lead acetate ($Pb(C_2H_3O_2)_2$) and monobasic calcium phosphate ($Ca(H_2PO_4)_2$). The borates form insoluble calcium borate salts, while the salts of tin (stannous) and lead form insoluble calcium stannite ($CaSnO_2$) and calcium plumbate ($CaPbO_2$), respectively. Others have proposed that insoluble hydroxides of the two heavy metals, $Pb(OH)_2$ and $Sn(OH)_2$, are formed and that they are responsible for the set retardation [3]. Whatever the nature of the insoluble product, it delays the normal hydration reactions. The water soluble monobasic calcium phosphate salt forms the very insoluble tribasic calcium phosphate $(Ca_3(PO_4)_2)$: Monobasic ammonium phosphate $(NH_4H_2PO_4)$ reacts similarly but liberates ammonia gas in the highly alkaline aqueous medium of the concrete. Both phosphate salts are the principal ingredients of commercial fertilizer. This is mentioned only because fly ash (to be discussed in Chapter 7) is occasionally transported in tank trucks that have previously contained fertilizer, becomes contaminated with the phosphate and when added to concrete as a mineral admixture, the phosphate acts to retard the time of setting of the resulting concrete. The insoluble salts coat the cement particles and delay their normal hydration processes.

ASTM TYPE D WATER REDUCING SET RETARDING ADMIXTURES

The Type B admixtures proved to be very effective retarders but sometimes unpredictable in the extent of their retarding action and were seldom used at addition rates exceeding 0.05% (solids) on weight of cement. The soluble compounds formed, or deposited, on the surface of the cement particles did not provide dispersion for those particles and, therefore, the amount of water reduction realized through their use was insignificant. Two other factors are responsible for their not being commonly used: cost and potential toxicological effects of waste waters from the wash-out of concrete mixing drums and admixture storage tanks.

The most commonly used chemical retarder is that of the Type D variety and is referred to as a water reducing and set retarding admixture [2]. The retarding ingredient which is organic in nature is adsorbed on the surface of the cement particles and imparts like and repelling charges to the surfaces. But, unlike the situation with water reducing admixtures (Chapter 3), a greater quantity of the dispersant is added, and more of the cement particle surface is shielded from their usual reactions with water. The water reducing set retarder also differs from the plain water reducing admixture in that it contains no supplemental set accelerating component. The dispersing components of the Type D admixtures are basically the same as those used in the

Type A additives, as shown in Table 5-1. The lower addition rate of the dispersing agent, coupled with small amounts of a set accelerator, in the Type A admixture is designed to give maximum dispersion with a minimum of set retardation, as depicted in Figure 5-1.

Composition

The formulation of a Type A admixture with the goal of producing maximum dispersion (or water reduction) with a minimum of delay in time of setting becomes a rather delicate balance between the amounts of its two components (dispersant and accelerator) and the overall addition rate of the combination. In the case of the Type D, the admixture must be designed to produce maximum water reduction and yet not exceed the maximum allowable extension of time of setting. Several variations in the formulae of Types A and D admixtures based on calcium lignosulfonate and their effects on water reduction and time of setting of concrete are illustrated in Table 5-2.

The data in Table 5-2 warrant some comment. First, all of the test concretes had a C.F. of 517 lbs (type 1), were batched at 72°F and R.H. of 50% and their times of setting were measured under those same environmental conditions. Second, consider the following results obtained for the five admixture treated concretes, with respect to the plain or reference concrete:

1. The Type A-1 formulation can be discarded because it does not produce sufficient water reduction.
2. When the amount of dispersant is doubled in Type A-2 (over that in Type A-1), the water reduction increases by a factor close to two, but the delay in time of setting, both initial and final,

Table 5-1.   A Comparison of the Usual Addition Rates of the Retarding-Dispersing Components of Types A and D Admixtures.

	ADDITION RATE—%[a]	
RETARDING-DISPERSING COMPONENT	TYPE A	TYPE D
Salts of lignosulfonic acid	0.16	0.22
Salts of hydroxylated carboxylic acids	0.04	0.06
Glucose polymers	0.04	0.06

[a]% Solids on weight of cement.

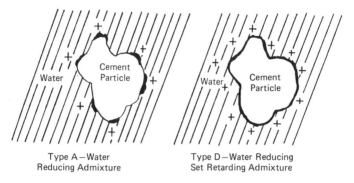

Type A—Water   Type D—Water Reducing
Reducing Admixture   Set Retarding Admixture

Figure 5-1. Effect of concentration of dispersant on the degree of cement particle surface coating.

exceeds the permissible maximum value of 90 minutes. So, it too can be eliminated.

3. The amount of dispersant added via formulation Type A-3 is the same as that introduced by Type A-2, and the water reduction caused by the two is the same. However, by doubling the amount of accelerator (vs. Type A-2) the delay in time of setting is now reduced to a point that it meets the set retardation requirements. The Type A-3 formulation meets the origninal goals, i.e., produces maximum water reduction with a minimal effect on time of setting. If it performs in a similar way in concrete made from a number of other portland cements, a determination of its effects on the other properties of cement is next in order.

4. The Type D-1 formulation produces satisfactory water reduction and degree of set retardation, but the magnitude of the latter indicates that its addition rates might be increased to provide more dispersion.

5. The amount of dispersant is increased in Type D-2 (over that in Type D-1), and the water reduction that it provides increased by almost two percentage points. The delay in time of setting still does not exceed the maximum allowable limit, but is approaching it. Because of the differences in composition and physical properties of portland cements, it is wise, at this point, to evaluate Type D-2 in concrete made from a number of different cements.

Table 5-2. Effects of Various Formulations of Type A and D Admixtures on Certain Properties of Concrete.

| ADMIXTURE COMPONENT | PLAIN | ADDITION RATE—% SOLIDS ON CEMENT | | | | |
		TYPE A-1	TYPE A-2	TYPE A-3	TYPE D-1	TYPE D-2
Calcium lignosulfonate	—	0.08	0.16	0.16	0.16	0.22
Triethanolamine	—	0.01	0.01	0.02	—	—
Concrete Properties						
Slump-in.	3-1/4	3	3-1/4	3-1/4	3	3
W-C ratio	0.59	0.57	0.54	0.54	0.54	0.53
Water reduction—%	—	3.4	8.5	8.5	8.5	10.2
Time of setting—min[a]						
Initial	210	235	335	255	310	365
Final	250	285	375	320	355	435
ASTM C494 [2]						
Water reduction—%	—	5 (min)				
Delay in time of setting—min						
Initial	—	90 (max)			210 (max)	
Final	—	90 (max)			210 (max)	

aReference [4].

Influence on C₃S Hydration

Typical isothermal calorimeter curves of a type 1 portland cement paste, with and without a Type D chemical admixture, are illustrated in Figure 5-2. A number of differences exist between the two curves.

1. The second exothermic peak (caused by the C_3S phase) is higher for the untreated paste.
2. The dormant period (discussed in Chapter 1) of the paste treated with the admixture is longer.
3. The areas under the two curves at the end of 24 hours (not shown) are essentially the same.

Both pastes have a W/C of 0.70.

Influence on Compressive Strength

The influence of the water reduction realized from the use of a Type D retarder and the absence of water reduction resulting from the addition of a Type B admixture is reflected in the relative compressive strength requirements of the treated concretes. For example, the 3-day and up to 1-year compressive strengths of concrete containing a

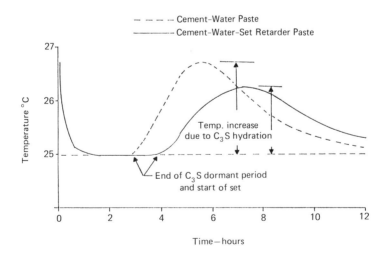

Figure 5-2. Isothermal conduction calorimetric curves of a Type I Portland Cement paste with and without a set retarding admixture.

Type B admixture must be at least 90% that of the reference (untreated) concrete, while that of concrete treated with a Type D retarder must be at least 110% of the reference at 3-, 7-, and 28-days of age and equal to the reference at 6 months and 1 year [2]. In actual practice, the compressive strengths of concrete treated with either of the types of retarding admixtures usually exceed these requirements by a considerable margin.

When one considers the 28-day compressive strength produced in concrete as a result of using a Type D admixture with that created by a set accelerating admixture (Type C), the former should be greater, if by no other reason, through a reduction in W/C. The author compared the 28-day compressive strength contribution of a Type D with that of a Type C, using the Omega Index Factor (O.I.F.) concept, described in Chapter 2, wherein any difference in W/C is taken into account as a part of the O.I.F. (abscissa) in Figure 5-3. The pertinent concrete data used to generate Figure 5-3 are summarized in Table 5-3. The increase in strength produced by the Type C, over the reference base line is 2% while that generated by the Type D is 12%. If the water reducing set retarding admixture was relying only on water reduction to increase strength, its concrete data point would fall on, or very close to, the reference base line. Therefore, the Type D admixture must be affecting strength (in a positive way) through some other mechanism.

The author pointed out in Chapter 4 that the presence of a set accelerating admixture causes more hydration products (of the C_3S) to be formed earlier, but of a smaller particle size, thus giving rise to higher early strengths and that the penalty for this phenomenon was reflected in lower later strengths. In fact, he showed that as the degree of set acceleration increased, the improvement in 28-day compressive strength (over the reference) decreased linearly (see Figure 4-14). It follows then that the degree of set retardation should also be related to the increase in 28-day strength, over the reference, but in a direct proportional manner.

In order to test this concept, two series of concretes were fabricated using a type I portland cement, a C.F. of 517 lbs, and a constant W/C of 0.56. The first series was batched and tested for time of setting at 73° F and 50% R.H. The cylinders that were cast for 28-day compressive strength measurements were moist cured at 72° F. One of the concretes was mixed with no admixture added (reference). The

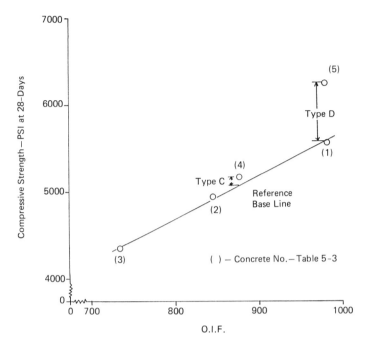

Figure 5-3. A comparison of the compressive strength contributions by Type C (accelerating) and Type D (water reducing set retarding) admixture at 28 days.

second, third, and fourth batches were treated with 0.03%, 0.06%, and 0.12% (solids on cement), respectively, with a Type D admixture based on the sodium salt of a hydroxylated carboxylic acid. In the second series, the concrete raw materials, mixing drum and cylinder molds were conditioned at 90° F (50% R.H.) for 48 hours. Four concretes were also prepared in this series, in the conditioning room, all treated with the same Type D admixture and at the same three addition rates just mentioned; i.e., 0.03%, 0.06%, 0.12%, and at 0.18%. The time of setting was measured at 90° F (50% R.H.), and the test cylinders were cured at 90° F in saturated lime water which had been preconditioned to 90° F. The results of that work are illustrated in Figure 5-4. The increase in compressive strength (ordinate) and increase in final time of setting (abscissa) in Figure 5-4 are based on those particular properties of the 72° F reference concrete.

   The concretes in the first series (72° F) exhibit a direct linear relationship between final set retardation and increase in 28-day com-

Table 5-3. Concretes Designed to Compare the Compressive Strengths Contributed by a Type C and D Chemical Admixture to Concrete, at 28-days, Using the O.I.F. Concept.

CONCRETE COMPONENTS AND PROPERTIES	CONCRETE NO.				
	1	2	3	4	5
Cement—lb/yd³ᵃ	540	490	440	500	500
Coarse Agg.—lb/yd³	1680	1725	1775	1715	1745
Fine Agg.—lb/yd³	1370	1405	1445	1400	1425
Water—lb/yd³	297	284	264	285	255
Admixture	—	—	—	Yesᵇ	Yesᶜ
Air—%	1.6	1.5	1.5	1.6	2.1
Slump—in.	3-1/2	3-1/4	3-1/2	3-1/4	3-1/2
W/C	0.55	0.58	0.60	0.57	0.51
O.I.F.	982	845	733	877	980
Compressive Strength at 28-days-psi	5560	4950	4355	5175	6250

ᵃType I portland cement.
ᵇ2% anhydrous CaCl₂ on weight of cement.
ᶜ0.22% calcium lignosulfonate on weight of cement.

pressive strength. In the 90° F series, the lowest addition rate of the retarder failed to overcome some of the acceleration caused by the elevated temperature resulting in a 25 minute decrease in the final time of setting and a 28-day strength gain of 105%, but at all of the other addition rates exerted a lesser retarding effect, as might be expected. However the data points essentially fall along the same line as those for the 72° F series. In the 90° F series there seems to be a "tug-of-war" going on; the elevated temperature wants to accelerate the time of set, and the retarder wants to delay the set.

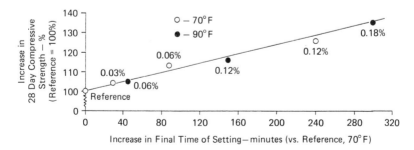

Figure 5-4. Influence of set retardation on compressive strength.

One can conclude from the data in Figure 5-4 that the slower formation of the concrete hydration products, due to the presence of a
set retarding admixture, is responsible for the higher later strengths.
As a result of using set retarding admixtures (Types B or D), the
normal hydration processes are slowed down so that the very early
strengths; i.e., at 1-day, are somewhat lowered because of the reduced amount of hydration products. At the same time, the slower
rates of formation of the hydration products gives them a greater opportunity to align, or organize themselves, in the paste component
and produce higher later strengths. This is somewhat analogous to the
construction of a brick wall. When the bricks are carefully mortared
into place, which takes time, the wall will be stronger than one in
which the bricks have been placed in a rapid, but haphazard, fashion.
The author likes to think of the hydration products of the silicates,
which are mainly responsible for strength, as undergoing a polymerization reaction, as have others [5]. As is the case with other polymerization reactions, the slower they proceed, the greater is the structural integrity of the finished product.

DELAYED ADDITION OF ASTM TYPE D ADMIXTURES

The question frequently arises as to whether the dispersing-set retarding chemical in the admixture is preferentially adsorbed on the
surface of the cement or on that of the initial hydration products. A
clue to the proper response to this question was brought to light in
the early 1960s when the author observed, as well as others, that the
time of addition of water reducing-set retarders exerted a considerable
effect on the time of setting of portland cement pastes and concrete,
as well as other properties of the cementitious mass [6][7][8]. Usually, the admixture is added to the concrete with the mix water, but
when it is introduced 2 or more minutes after all of the other ingredients have been added and mixing has begun, the time of setting is
delayed an additional 2 or 3 hours. This technique came to be known
as "delayed addition" in the years that followed and was discussed
earlier in Chapter 3.

Some of the results of the authors work dealing with the delayed
addition phenomenon are shown in Tables 5-4 and 5-5. The cements
used are described in Table 5-4, and the properties of the concretes
fabricated from those cements are listed in Table 5-5. The water con-

Table 5-4. Compound Composition of Portland Cements Used in the Delayed Addition Experiments.

	APPROXIMATE COMPOSITION—%					
CEMENT NO.	C_3A	C_3S	C_2S	C_4AF	TOTAL ALKALI[a]	SO_3
A	10.1	46.8	25.2	8.2	0.71	2.81
B	11.6	48.2	24.3	7.0	0.93	2.14
C	5.9	49.5	28.1	9.3	0.48	1.76
D	9.6	44.5	28.4	6.1	0.91	2.80
E	14.5	43.0	25.8	7.7	0.53	2.70

[a]Total alkali, expressed as Na_2O

tents of the concretes made from cements B, C, and D were adjusted to produce essentially the same slump, because one of the results of the delayed addition is an increase in this property of concrete. An increase in the air content of the treated concrete is another result. Whether or not the two effects are related is an argument that is discussed in Chapter 6. The important point here concerns the influence

Table 5-5. Physical Properties of Concrete with and without the Delayed Addition of Water Reducing-Set Retarding Admixture.

					DELAY IN TIME OF SETTING-MIN	
CEMENT NO[a]	METHOD ADDITION[b]	W/C	SLUMP-IN.	AIR-%	INITIAL	FINAL
A	1[c]	0.54	4	1.0	—	—
	2	0.48	4	3.6	60	85
	3	0.48	8	5.5	235	240
B	1	0.59	4	2.2	—	—
	2	0.55	3-1/2	5.1	90	105
	3	0.51	3-1/4	6.9	240	270
C	1	0.53	3-3/4	1.0	—	—
	2	0.48	3-3/4	3.1	134	140
	3	0.48	4	3.3	250	255
D	1	0.58	4-1/2	2.5	—	—
	2	0.54	3-1/2	4.6	65	115
	3	0.50	4-1/2	6.7	215	285
E	1	0.53	4	1.2	—	—
	2	0.49	3-3/4	3.5	60	60
	3	0.49	7	5.9	180	170

[a]From Table 5-3, C.F. = 564 lb.
[b]Addition rate of calcium lignosulfonate = 0.225% solids on weight of cement.
[c]1—none, 2—addition with mix water, 3—added after 2 minutes of mixing.

of the delayed addition on the degree of retardation of initial and final times of setting.

The results of the work illustrated in Tables 5-4 and 5-5, along with those more recently generated, makes the author inclined to agree with Bruere's thinking [6].

- When a water reducing-set retarding admixture is added to concrete with the mix water, it is adsorbed on the aluminate phases of the cement before any appreciable amount of calcium sulfate can dissolve in the aqueous phase and make itself available for reaction with those phases. This leaves less dispersant available for adsorption on the silicates and retard their hydration reactions.
- The delayed addition of the admixture allows the calcium sulfate time to dissolve in the aqueous phase and coat the aluminates with ettringite. When the admixture is finally added, the aluminates are unable to adsorb it, and a larger than normal amount of retarder becomes available to delay the silicate hydration reactions.

The author also adds to Bruere's concept that the larger amount of available dispersing retarder, resulting from its delayed addition, can act to disperse, in addition to the cement, the very fine particles introduced by the aggregates and the entrapped air in the mix, the latter resulting in an increase in the air content of the concrete. The results of the work of other investigators tend to support these concepts, although some controversy exists [9][10][11][12][13].

The concept just presented is somewhat substantiated by the results obtained from six concretes fabricated from two cements. Cement No. 1 was made by intergrinding a type I clinker with sufficient gypsum to produce a final product having 2.5% SO_3. Cement No. 2 was prepared by intergrinding the same clinker with natural anhydrite to give the finished cement an identical SO_3 content. Both cements had a Blaine fineness of approximately 3450 cm^2/g. Concretes were then fabricated from the two cements. In one series, no admixture was added. In the second series, a Type D admixture was added to the test concretes after they had mixed for 2 minutes. The properties of the concretes are summarized in Table 5-6.

The data in Table 5-6 warrant comment. With respect to the untreated concretes, (1) and (4):

Table 5-6. Properties of Concretes Made from Type I Cements Containing Gypsum and Natural Anhydrite, with and without a Type D Admixture.

CONCRETE PROPERTIES	CEMENT NO.[a]	NO. ADMIXTURE (1)	ADMIXTURE ADDITION[b]	
			WITH MIX WATER (2)	2 MIN. DELAY (3)
W/C	1	0.58	0.53	0.53
Slump-in.	1	3-1/2	3-1/2	6-1/2
Air—%	1	1.5	2.0	5.8
Final Set—min.	1	245	395	465
	—	—	—	—
		(4)	(5)	(6)
W/C	2	0.58	0.53	0.53
Slump-in.	2	3	1-3/4	7
Air—%	2	1.4	1.6	5.6
Final Set—min.	2	210	55	410

[a]C.F. = 517 lb, materials, concrete and ambient temperature, 71 ± 1° F.
[b]Type D admixture, 0.25% calcium lignosulfonate on weight of cement.
Concrete number ( ).

- The concrete made from Cement No. 1 behaved normally.
- The concrete made from Cement No. 2 had a 35 minute shorter setting time than that made from Cement No. 1. This is most probably due to the low solubility of the natural anhydrite and thus permitting some direct reaction between the aluminates and water.
- Both concretes entrained the same amount of air and had essentially the same initial slumps.

As for the concretes treated with the admixture, added with the mix water, (2) and (5):

- The concrete made from Cement No. 1 behaved normally, exhibiting excellent water reduction (approximately 8%), a 150 minute extension of final time of setting and an increase in air content of only 0.5 percentage points.
- The concrete made from Cement No. 2 behaved abnormally. In addition to the poor water reduction (approximately 3%, based on slump—see Chapter 3), its time of final setting was actually accelerated by 155 minutes over that of (4), which is probably

due to the adsorption of the admixture by the natural anhydrite (see Chapter 3 where the influence of natural anhydrite on the properties of portland cement is discussed).

With regards to the concretes treated with the admixture after a 2 minute delay:

- The slump and air content of the concrete made from Cement No. 1 increased dramatically, over that of (2) even though the W/C was the same as that used when the admixture was added with the mix water. The increase in time of final setting was 220 minutes over the plain concrete and 70 minutes over that of the concrete treated with the admixture by way of the mix water. This falls within the trends set forth in Table 5-5.
- The slump and air content of the concrete containing Cement No. 2 (6) were essentially the same as those of the concrete made from Cement No. 1 (3) and its increase in final time of setting was 200 minutes over the plain concrete (4) and 355 minutes over that of the concrete treated by the admixture by way of the mix water.

All of this says that in order to get maximum set retardation from a Type D (or C) admixture, there must be a proper balance between the soluble SO_3 and aluminate phases of the portland cement. One other point should be made here and that has to do with field problems wherein a set retarding admixture is not performing as expected. Try the delayed addition technique; it could get the ready mix producer, as well as the contractor, out of trouble.

SLUMP LOSS OF CONCRETE CONTAINING SET RETARDING ADMIXTURES

The fabrication, placement and curing of portland cement concrete in hot weather is the principal application of set retarding admixtures whether they be of the Type B, D, or G. Hot weather is often considered to be any combination of high air temperature and low humidity and is aggravated by an increase in wind velocity. Because of its accelerating influence on time of setting, higher than normal temperatures will increase the rate at which the slump of concrete decreases, and this increases the contractor's tendency to add water at the job site. Without the addition of water, placing and curing along with cold joints loom as potential problems. In addition, controlling the entrained air content of the concrete often becomes a vexation.

Slump loss of plain concrete as a function of temperature was discussed briefly in Chapter 3 (Figure 3-20). In the presence of a set retarding admixture, the effect of temperature is modified significantly. The slumps of concretes fabricated and tested at 70°F and 90°F, with and without the addition of a Type D retarder are plotted as a function of time in Figure 5-5. All four of the laboratory test concretes were batched to the same initial slump of 4-1/4".

Since there is no established method for measuring the slump loss of fresh concrete, the author arbitrarily chose the procedure outlined in Table 5-7 for all of the work reported here and in Chapter 3. Fre-

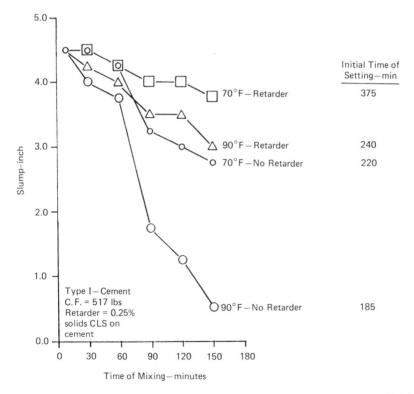

Figure 5-5. Influence of a Type D admixture on the loss in slump and time of initial set of concrete, at two temperatures.

Table 5-7. Laboratory Procedure for Measuring Concrete Slump Loss

STEP NO.	TIME PERIOD—MIN.	OPERATION
1	0–8	Mix [14]
2	8–11	Measure slump and return concrete to mixer (initial slump).
3	11–29	Mix at low speed with mouth of mixer covered.
4	29–31	Same as Step 2.
5	31–59	Same as Step 3.
6	59–61	Same as Step 2.
7	61–89	Same as Step 3.
8	89–91	Same as Step 2.

quently the test period was extended to cover longer periods of time but seldom exceeded 3 hours. In the experiments involving the concretes described in Figure 5-5, sufficient concrete was set aside from each mix, after the initial slump was measured, for determining time of setting [4]. In this instance, only the initial times of setting were of interest because the author was hoping to establish a relationship between the slumps of the concretes after prolonged mixing and their initial times of setting. A relatively smooth curve can be drawn through the data points, as illustrated in Figure 5-6. He did not expect to find a direct linear relationship between the two variables because the concrete in the mixer was subjected to essentially constant agitation while that in the time of setting test was stationary or dormant. In addition,

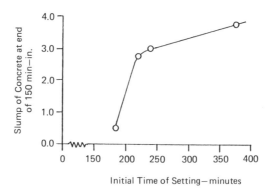

Figure 5-6. Relationship between slump—after prolonged mixing—and initial time of setting.

the different W/Cs used to attain the same initial slump also affects both the slump loss and time of setting. With regards to the slump loss values shown in Figure 5-5, the concrete at 90°F that contained the set retarding admixture exhibited slightly less (1/4") slump loss than did the plain concrete at 70°F and much less (2-1/2") than that of the untreated concrete at 90°F.

HOT WEATHER CONCRETING PRACTICES

Hot weather can also exert a number of negative effects on the properties of hardened concrete, such as (1) reduce later compressive strengths due to acceleration of time of setting coupled with the possible addition of extra water, (2) decrease in durability due to lack of control of the amount of entrained air, and (3) increase in tendency to undergo differential thermal cracking and drying shrinkage.

In Chapter 4 the author pointed out that any factor that decreases the time of setting of concrete usually lowers the compressive strength at later ages and that both of these changes become progressively magnified as the temperature of mixing, placing and curing increases. For example, it has been shown that concrete mixed, placed and cured at 70°F can have approximately 1000 psi greater compressive strength at 28 days than the same concrete when mixed, placed and cured at 100°F, at the same age [15].

The above normal temperatures of its surroundings in combination with the increase in rate of heat generated by its accelerating effect on the hydration of the cement leads to a "snow ball" influence on the time of setting and subsequent decrease in later strengths. There is not much one can do about the high temperature of the environment, if the contractor and/or engineer chooses not to protect the concrete, but there is something that can be done to control the rate at which the heat, produced by the hydrating cement, is generated—and that is through the use of a set retarding admixture.

ECONOMICS OF USING SET RETARDING ADMIXTURES

The influence of temperature on the set retarders capabilities of two commonly used Type D admixtures is illustrated in Figure 5-7. The ranges depicted in Figure 5-7 cover concretes made from various type I, II, and III cements at C.F.s of 500 to 520 lbs and at W/Cs of 0.55

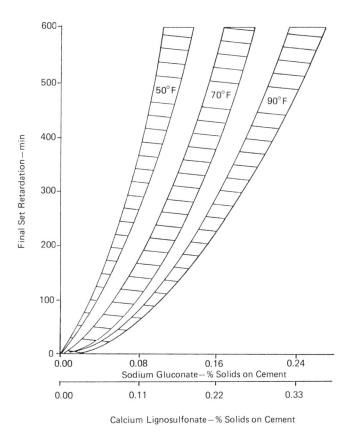

Figure 5-7. Influence of temperature on the time of final setting of concrete containing water reducing set retarders.

to 0.60. All of the concrete raw materials were pre-conditioned, mixed, and the resulting concretes tested at the stipulated temperature. It is quite evident from the data in Figure 5-7 that as the temperature of the concrete increases, the addition rate of the set retarding admixture must be increased to maintain the same degree of set retardation. Probably just as important is the lack of linearity, at a given temperature, between the amount of set retardation and the addition rate of the retarding admixture.

Of course, the contractor can specify that the concrete, as delivered to the job site, have a certain maximum temperature, say 80°F to 90°F. In order to meet this requirement during the hot summer months, the concrete producer must use cold mix water or a mixture of cold mix water and crushed ice as well as cool aggregates. The temperature of the cement in the storage silo can be as high as 150°F to 180°F, if it has been recently delivered, but there is essentially little that the concrete producer can do to lower its temperature. The use of cold water and cool aggregates will increase the cost of the concrete $2 to $4/yd³. Most of these precautions can be mollified when a set retarding admixture is added to the concrete, the cost of which will range from $1.50 to $1.75/yd³, depending on the admixture addition rate and the C.F. of the treated concrete.

Whatever the path taken by the concrete producer might be to provide concrete of acceptable temperature, the placing and curing precautions that are taken at the construction site are still very important to successful hot weather concreting. Such measures as (1) dampening the subgrade and forms, (2) erecting wind breakers and sunshades, and (3) reducing the time between placement and start of curing are frequently recommended. Curing should be aimed at either reducing the rate of evaporation of water from the concrete through the application of a concrete curing compound or covering with wet burlap or replacing the lost water by applying a mist or fog to the exposed surfaces. Both operations will minimize the loss in later strength and plastic shrinkage cracking. The latter is due to the rapid drying of the surface of fresh concrete after it is placed and is still plastic. These cracks must be closed by refinishing to assure adequate integrity of the hardened concrete.

While a great deal of evidence exists, derived from both laboratory and field investigations, that clearly shows that elevated mixing and curing temperatures lead to lower concrete strengths at 28 days, these observations have been made on concretes whose setting characteristics had not been altered by set retarding admixtures. The results of an extensive field test (1336 concretes, made from 7 different cements at 7 different C.F.s on 7 different construction projects) in which the effects of placing temperatures, ranging from 50°F to 99°F, on the compressive strengths of concretes treated with various set retarding admixtures, indicate that the admixture treated concretes exhibit strength characteristics at elevated placing and site curing temperatures that do not change from those of concretes used at lower temperatures [16]. These findings are in agreement with the results of laboratory research regarding the favorable effect of retarding admixtures on concrete performance at elevated temperatures [17][18].

Other major uses of set retarding admixtures include (1) mass concrete where it is important to restrain the heat liberated, on a heat liberated vs. time basis, of the hydrating cement to minimize thermal gradients within the mass and reduce the possibility of thermal cracking, and (2) in the construction of large bridge decks where the time of setting of successive concrete placements must be adjusted to occur simultaneously to avoid cold joints and dead-load stress cracking. In the latter application, the admixture must extend the plastic characteristics of the concrete until the progressive deflection resulting from increasing loads is complete.

USE OF SET RETARDERS IN STEAM CURED CONCRETE

The author has been discussing the potential deleterious effects of high temperature on the properties of hardened concrete. Now, consider the situation wherein the temperature of the concrete is purposely elevated, such as in the steam curing of precast and/or prestressed concrete products. The principal goal in this type of operation is to produce a product of sufficient strength to withstand the removal of casting forms and/or release of tension on the prestressed steel cables, within the concrete member, as soon as possible without appreciably affecting its later strength. The driving force behind such a program is the attainment of maximum re-use time of the casting forms (or beds) and a low in-process inventory, both of which are important economical factors. Over the past two decades, the use of set retarding admixtures has become common practice, especially in the warmer months in these particular concrete industries.

Steam, at atmospheric pressure, is the most commonly used source of heat energy, and a typical curing cycle consists of four steps.

Step 1. Presteaming Period (often referred to as Preset Period). The concrete is fabricated at the precast plant and immediately transferred to and placed in the forms. The form and its contents are completely enclosed and allowed to remain unmolested for at least 3 hours but rarely longer than 7 hours. During this time, depending upon the temperature of the concrete, its W/C, C.F., and mass as well as shape, the concrete undergoes initial set. If the period is reduced to less than 3 hours, and before initial set has occurred, the early as well as later compressive strengths will be reduced. During this period the temperature of the concrete rises because of the heat generated within the mass by the hydrating cement and because that heat is restrained from escaping by its enclosure. When to end this period is not easily determined, even when a sample of the concrete is stored next to the precast member within the enclosure for measurement of its time of initial setting.

Step 2. Temperature Rise Period. Steam is introduced into the concrete enclosure at such a rate that the temperature within the enclosure rises at a maximum rate of 40°F/hr to 60°F/hr until a maximum temperature of 135°F to 165°F is reached. The temperature ranges just cited are necessarily broad because every producer operates with a heating schedule that works best with the concrete raw materials with a view toward quality of finished product and economics (steam costs money). If the rate of temperature rise is too rapid, the temperature differential between the concrete and its surroundings can lead to stresses that promote cracking.

Step 3. Steaming Period. After the chosen maximum temperature is reached, it is maintained for a period of 6 to 12 hours. The duration of steaming is based upon the required compressive strength at the end of the completed cycle and at later ages. When the steam is first introduced (Step 2), the temperature of the concrete slowly reaches that of the steam in the enclosure, then surpasses it due to the cement's heat of hydration. When the temperature of the concrete and that of the envelope atmosphere are the same, additional steaming may cause the vapor pressure in the concrete to exceed that of its surrounding atmosphere and cause drying, shrinkage, and a loss of strength [19].

Step 4. Cooling Period. At the end of Step 3, the temperature of the concrete is reduced to that of the precast plant at a regulated rate, i.e., 40°F/hr to 70°F/hr. Rapid cooling can cause cracking and crazing due to the formation of wide temperature gradients and differential stresses in the concrete member.

One reason, and probably the principal one, that the presteaming period is included in the steam curing cycle lies in the nature of the early hydration reactions of the cement. The author pointed out in Chapter 1 that as soon as portland cement comes in contact with water, the aluminate phases, and in particular the tricalcium aluminate, starts to react with it and calcium sulfate to form calcium sulofaluminates. In the absence of calcium sulfate (gypsum) or when an insufficient amount of it is present in the aqueous phase, the aluminates in the cement react directly with water to form aluminate hydrates. The presence of the hydrates lowers the strength of the hardened concrete and makes it very susceptible to sulfate attack.

The solubility of calcium sulfate (gypsum) in water, unlike most inorganic salts, decreases markedly and almost linearly with an increase in temperature. For example, its solubility in water at 160°F is only about one-third that at 70°F. The presteam period allows the maximum amount of the sulfate to go into solution in the concrete aqueous phase, early in its life, and react with the aluminate phases, thus minimizing the chances for aluminate hydrate formation. If the presteaming period were to be eliminated, or drastically reduced, the increase in solubility of the calcium sulfate during the cooling period makes it now available to react with the aluminate hydrates causing disruptive expansion much like the sulfate attack phenomenon described in Chapter 7.

When a set retarding admixture is used in the steam curing operation, a push-pull situation is created within the concrete. The purposeful increase in temperature of the concrete in Step 2 and in Step 3 acts to accelerate the hydration of the tricalcium silicate phase of the cement which normally leads to high early compressive strengths but lower later strengths (Chapter 4). In the presence of a set retarding admixture, the rate of silicate hydration is reduced, in spite of the increased temperature, off-setting most of the normally expected later strengths (Figure 5-4). One might expect that the retarder would counteract some of the early strength gains realized through the temperature increase, but very little, if any, is noted. This is probably due to the slower evolution of the heat of hydration during Step 2, thus minimizing the chances of the temperature of the concrete exceeding that of the atmosphere in its enclosure during Step 3. The optimum addition rate of the admixture is usually determined by trial, as is the optimum steam curing cycle (Steps 1-4). The use of a water reducing set retarding admixture also provides an extra boost in strength, at all ages, by making the use of a lower W/C in the concrete possible.

ASTM TYPE F, HIGH RANGE WATER REDUCING SET RETARDING ADMIXTURES

The Type G admixture is referred to as a water reducing, high range retarder. Although its stipulated influence on the time of setting of concrete is the same as that for a Type D admixture, its required effect on W/C is greater as is its improvement of compressive strength of

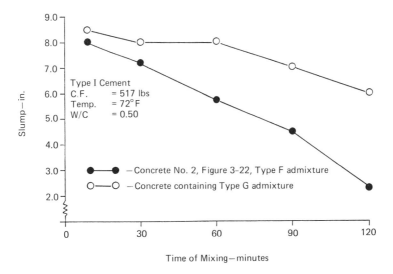

Figure 5-8. Comparison of concrete slump loss in the presence of ASTM Type F and G admixtures.

concrete at ages of less than 28 days [2]. A Type G is basically a Type F but contains an added set retarding component.

One of the objections raised concerning the use of the Type F admixture is its tendency to aggravate the slump loss of the treated concrete. A typical example of that behavior is shown in Figure 3-22 (Concrete No. 2), where the slump of the treated concrete drops 5-3/4" over a mixing period of 120 minutes. The slumps of concretes made from a type I cement and treated with a Type F and with a Type G admixture are compared, as a function of time of mixing, in Figure 5-8. The loss in slump of the concrete containing the Type G only decreased 2-1/2" over the period of test, about one-half of the decrease noted for the Type F concrete. Both admixtures contained a naphthalene based high range water reducing compound, but the Type G contained a small amount of a salt of a hydroxylated carboxylic acid as the retarding component and both were added at the same addition rate, i.e., 0.40% solids on weight of cement.

REFERENCES

[1] ACI 116, "Cement and Concrete Terminology," *American Concrete Institute, SP 19*, Detroit, MI, pg. 117 (1988).

[2] ASTM C494, "Standard Specification for Chemical Admixtures for Concrete," *Annual Book of ASTM Standards*, Vol. 04.02, pp. 245-252 (1988).

[3] Lieber, W., "The Influence of Lead and Zinc Compounds on the Hydration of Portland Cements," *Proceedings 5th International Symposium on the Chemistry of Cements*, Tokyo, Vol. 2, pp. 444-453 (1968).

[4] ASTM C403, "Standard Test Method for Time of Setting of Concrete Mixtures by Penetration Resistance," *Annual Book of ASTM Standards*, Vol. 04.02, pp. 210-213 (1988).

[5] Daugherty, K.E., Skalny, J., "The Slowest Polymerization Reaction," *Chemistry*, Vol. 45, No. 1, pp. 12-15 (1972).

[6] Bruere, G. M., "Importance of Mixing Sequence When Using Set-Retarding Agents with Portland Cement," *Nature*, Vol. 199, pg. 32 (1963).

[7] Cook, H.K., "Two Factors Affecting Results of ASTM Method C403; (1) Time of Addition of Admixture; (2) Use of Mortars in Lieu of Concrete," *Highway Research Board Meeting*, Washington, DC, Jan. (1963).

[8] Dodson, V.H., Farkas, E., "Delayed Addition of Set Retarding Admixtures to Portland Cement Concrete," *Proceedings, American Society for Testing and Materials*, Vol. 64, pp. 816-826 (1965).

[9] Blank, B., Rossington, D.R., Weinland, L.A., "Adsorption of Admixtures on Portland Cement," *Journal of the American Ceramic Society*, Vol. 46, No. 8, pp. 395-399 (1963).

[10] Young, J. F., "A Review of the Mechanisms of Set-Retardation in Portland Cement Pastes Containing Organic Admixtures," *Cement and Concrete Research*, Vol. 2, pp. 415-433 (1972).

[11] Ramachandran, V.S., "Interaction of Calcium Lignosulfonate with Tricalcium Silicate, Hydrated Tricalcium Silicate and Calcium Hydroxide," *Cement and Concrete Reasearch*, Vol. 2, No. 2, pp. 179-194 (1972).

[12] Monosi, S., Moriconi, E., Pauri, M., Collepardi, M., "Influence of Lignosulfonate, Glucose and Gluconate on the C_3A Hydration," *Cement and Concrete Research*, Vol. 13, pp. 568-574 (1983).

[13] Collepardi, M., Monosi, S., Moriconi, G., Pauri, M., "Influence of Gluconate, Lignosulfonate and Glucose Admixtures on the Hydration of Tetracalcium Aluminoferrite in the Presence of Gypsum With or Without Calcium Hydroxide," *Journal of the American Ceramic Society*, Vol. 68, pp. 126-128 (1985).

[14] ASTM C192, "Standard Practice for Making and Curing Concrete Test Specimens in the Laboratory," *Annual Book of ASTM Standards*, Vol. 04.02, pp. 110-116 (1988).

[15] ACI Commitee 305, "Hot Weather Concreting, ACI 305R-77," *American Concrete Institute Proceedings*, Vol. 74, No. 8, pp. 324-325 (1977).

[16] Mittelacher, M., "Effect of Hot Weather Conditions on the Strength Performance of Set-Retarded Field Concrete," *Temperature Effects on Concrete, ASTM STP 858*, American Society for Testing and Materials, Philadelphia, pp. 88-106 (1985).

[17] Tuthill, L.H., Cordon, W.A., "Properties and Uses of Initially Retarded Concrete," *American Concrete Institute, Proceedings*, Vol. 52, No. 3, pg. 282 (1955).

[18] Berge, D., "Improving the Properties of Hot-Mixed Concrete Using Retarding Admixtures," *American Concrete Institute, Proceedings*, Vol. 73, No. 7, pg. 396 (1976).
[19] Schmid, E., Schutz, R.J., "Steam Curing," *Journal, Prestressed Concrete Institute*, Vol. 2, No. 2, pp. 37-40 (1957).

Chapter 6

AIR ENTRAINING ADMIXTURES

HISTORY AND GENERAL COMPOSITION

While the early history of the use of air entraining admixtures is ambiguous, the author has heard, but not read, that the early Romans and Greeks added them to their pozzolanic mixes to increase their workability. Probably blood or animal fat was used for this purpose. Before proceeding further with the evolution of this class of admixtures, it should be defined in acceptable and modern terms. An air entraining admixture is simply one that is added to either portland cement paste, mortar, or concrete for the purpose of entraining air in the respective masses. As you will read later in this chapter, the entrainment of air in concrete has a number of beneficial effects, when used properly, on certain of its properties. The most important of these is an increase in resistance to frost attack and to deterioration, due to exposure to repeated freezing and thawing.

Up until the mid-1930s, the beneficial effects of air entrainment had been noted, but it was not known that the air was responsible. The first mention of air entrainment as a possible cause for improved frost resistance of concrete was made in 1939 [1]. Since the results of that same work also indicated that entrained air reduced bleeding and improved the uniformity of the concrete, the frost resistance caused by air entrainment was considered an indirect rather than a direct one.

Concretes made from cements that had been interground with small amounts of certain organic materials were also found to exhibit an increase in resistance to frost attack [2]. A second study, reported in the same year, found that the presence of certain organic compounds decreased the detrimental effect of freezing and thawing whether de-icing salts, such as calcium chloride or sodium chloride, were used to melt snow or ice on pavements or not used [3]. During the years that followed, it became increasingly accepted that the frost resistance and freeze-thaw durability caused by entrained air was its primary function, and the decrease in bleeding was a secondary result.

The development of methods for measuring the air content of freshly mixed concrete was soon to follow. The first was the gravimetric method, then the volumetric method and finally the pressure method. All three have withstood the test of time but with some modifications and are presently accepted methods by the industry [4][5][6]. The next question the industry had to address concerned the measurement of the amount of air, its distribution, the size of the entrained air void in hardened concrete and finally the extent to which these parameters influenced its resistance to frost and freezing and thawing. One of the first methods to be developed was that of linear traverse [7]. Since its introduction, it has been altered, modified, and refined, but the original principles have suvived and it has become a standard practice [8].

From the data generated during the development of the linear traverse method and that collected soon after its introduction it was concluded that the size, distribution, and other characteristics of the entrained air void were of more importance than the total volume of entrained air to the concrete's resistance to frost, freezing, and thawing [9][10]. These principles are still considered valid.

Specifications for materials proposed for use as air entraining admixtures (hereafter referred to as AEAs) in concrete and the accepted method by which they are evaluated have been established by ASTM [11][12]. While the earliest AEAs were animal tallow or complex hydrocarbons (oils/greases) which had been oxidized during the manufacture of cement, and inadvertently found their way into concrete, the modern AEAs are more sophisticated in nature and are purposely added to concrete to produce the desired beneficial effects. Most of the modern AEAs are anionic in character because of the stability that they impart to the entrained air void. Cationic AEAs have been proposed for use, but their cost is prohibitive, and the stability that they provide to the air void is questionable. A few non-ionic materials, such as water soluble low molecular weight ethylene oxide polymers, are being used, but they too offer very little stability to the entrained air void.

MECHANISM OF AIR ENTRAINMENT

The anionic type includes the water soluble salts (usually sodium) of wood resins, wood rosins, lignosulfonic acid, sulfonated hydrocarbons, proteinaceous materials, and fatty acids. Most of these are by-products of the paper industry, refinement of petroleum or processing of animal fat and hides. Most recently, a methylester-derived cocamide diethanolamine has been introduced which is claimed to create a more uniform size of entrained air bubble of greater stability in concrete [13].

There are a number of misconceptions about the process through which AEAs act that must be clarified. First, an AEA *does not* generate air in the concrete but merely stabilizes the air either (1) infolded and mechanically enveloped during mixing, (2) dissolved in the mix water, (3) originally present in the intergranular spaces in the dry cement and aggregate, or (4) in the pores of the aggregate. Second, while it is true that the entrained air is within the total mass of concrete, it is only entrained in the paste portion of the mix. Of course, when excessive amounts of air are entrained, some of the air voids will collect at the paste-aggregate interface, but, in general, when the proper amount of air is entrained, it will be in the paste portion of the concrete.

In the absence of admixtures all concretes contain some air as a result of the same factors just cited. However, most of that air is what is often referred to as "entrapped" air. In order to better appreciate the difference between entrained and entrapped air voids, the definition of the two must be considered. With respect to the former, it is considered to be a space in cement paste, filled with air whose diameter is typically between 10 μm (1×10^{-3} in.) and 1 mm (0.039 in.) and essentially spherical in shape. The latter type void is characteristically 1 mm (0.039 in.) or more in diameter and irregular in shape [14]. In air entrained concrete that contains 5% to 6% air, after placing, the *paste* will contain 7 to 24 million air voids/in.^{3} [15].

The air produced (mostly entrapped) in mortar by a given portland cement, in the absence of an admixture, is measured in accordance with ASTM C185 and usually reported in the producer's cement mill certificate [16]. The ASTM C185 test involves mixing 350 grams of the test cement, 1400 grams of a specially graded sand and enough water to produce a mortar whose flow (or workability) is within a designated range. A portion of the mortar (400 mL) is then weighed, and its air content is calculated from its weight and volume. These values vary from cement to cement and from the same type and manufacturer. The author recently found that the amount of air produced by 47 different cements, of various types and brands, averaged 8.8%, but the values ranged from 6.5% to 10.4%. If the air contents of another 47 cements had been analyzed, the chances of the average and the extent of the range being different would be good.

In order to relate the air produced by the cement in plain mortar (ASTM C185) and that produced by the same cement in plain concrete, one must perform a number of mathematical manipulations. Consider, for example, the concrete described in Table 6-1. The volume of air entrained/entrapped in the *concrete* is 0.46 ft^3/yd^3 of concrete and that of the *paste* in that cubic yard of concrete is 8.11 ft^3. Therefore, the amount of air contained by the *paste* in the concrete is 6% (0.46 ÷ 8.11). The amount of air produced by the same cement in mortar by the ASTM C185 method was found to be 9.2%. It can be calculated that the *paste* in the mortar contained approximately 30% air, by volume, five times that in the concrete *paste*. This difference is basically due to the difference in the mixing of the two final products and their sand gradation. As a general rule, and the author calls it Dodson's Rule No. 3, the concrete producer can roughly estimate the amount of air that will be entrapped/entrained in the plain concrete by dividing the air content reported by the cement producer on a mill certificate by a factor of *five*.

Now consider the thermodynamics of air void coalescence in plain water.

$$P = 4\,\gamma/d \qquad\qquad (6\text{-}1)$$

Table 6-1. Properties of Plain Concrete.

CONCRETE COMPONENTS[a]		CONCRETE PROPERTIES
Cement	515	Slump-in. 5-1/2
Fine Agg.	1436	Air − % 1.7[b]
Coarse Agg.	1756	Yield-ft^3/yd^3 27.1
Water	314	

[a]lb/yd^3
[b]Reference [6]

According to Equation 6-1, as the diameter of the air bubble is de-creased, the air pressure within the bubble increases. Equation 6-2 shows that any increase in pressure within the air void will increase the solubility of the air in the aqueous phase surrounding the bubble and thus make the void even smaller.

$$C = KP \qquad (6\text{-}2)$$

where
   P = pressure within the air void
   γ = surface tension of water
   d = diameter of air void
   C = concentration of air in water
   K = constant

This process of air dissolution will continue until some of the bub-bles disappear completely. The loss of air by the smaller bubbles causes an equivalent amount of air to come out of solution and enter the larger bubbles where the pressure is less. The net result is that the larger bubbles get larger at the expense of the smaller ones. This same phenomenon is probably occurring in the aqueous phase of plain non-air entrained concrete.

When AEAs are added to concrete, they form a film at the air void-water phase interface such as that shown in Figure 6-1, and their films can vary in their solubility in the aqueous phase. The hydrophilic groups in the anionic type combine with the calcium ion in the aqueous phase and form insoluble calcium salts. Generally, the hydrophilic groups in the cationic AEAs form hydroxides of varying solubilities.

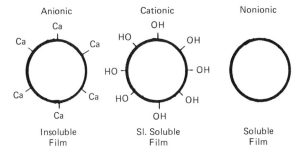

Figure 6-1. Nature of the AEA film around the entrained air void.

The non-ionic AEAs are essentially chemically inert; i.e., do not combine with the ions in the concrete aqueous phase, and thus remain soluble. An attempt to illustrate this difference is shown in Figure 6-1. If the coatings are sufficiently insoluble, the normal thermodynamics of coalescence are impeded, and the tendency of the small bubbles to grow smaller, and the larger bubbles to grow bigger is minimized.

Since the common AEAs vary widely in composition, a variation in their effectiveness might be expected. Many criteria have been set by others regarding their capacity to entrain air, but the author has given the following top priority:

1. The AEA must form a film around the air void of sufficient elasticity to resist internal and external pressures in its environment.
2. The AEA must form a film that resists deterioration with time and thus inhibit coalescence through transmission of air across the air-AEA coating-water interface.
3. The entrained air does not impair the properties of the concrete.

One of the important characteristics of the entrained air void in hardened concrete is its chord length. Why the chord length rather than the diameter? Consider filling a large box with ping-pong balls and cutting the box in half, either vertically or horizontally. The chances of dissecting one or more balls at its diameter is miniscule. An analogous situation arises when a sample of hardened concrete is cut (and polished) prior to examination by the method of linear traverse. Therefore, the chord length of the air void must be considered a measure of its size. The chord length distribution of the air voids entrained in concrete by representatives of the three classes of AEAs is shown in histogram form in Figure 6-2. The three concretes from which that data were derived had a C.F. of 611 lb and a freshly mixed air content of 5-1/2 ±1/4%. All of the chord lengths greater than 450 μm have been lumped together at the right hand side of each histogram. It is evident that there is a considerable difference in distribution. The non-ionic AEA gives rise to a larger bubble size while the anionic AEA produces bubbles, the greater percentage of which is in the smaller range. The cationic AEA produces air voids whose size ranges somewhere between these two extremes.

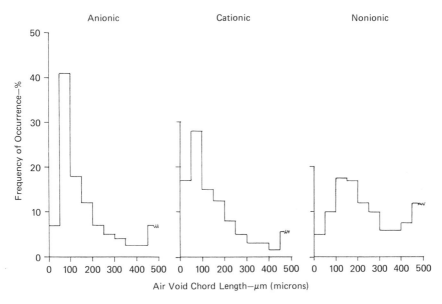

Figure 6-2. Histogramic display of the air void chord lengths of those entrained by the three types of AEAs.

The AEAs based on the alkali salts of wood rosins and resins react almost immediately with the calcium in the aqueous phase of the concrete to form insoluble calcium resinates or rosinates. This can be easily demonstrated by adding a few drops of the AEA to a small amount of saturated lime water. A colloidal precipitate immediately forms followed by rapid coagulation and finally precipitation in the nature of large curds. When the same experiment is performed using an anionic AEA based on a soluble tall oil or sulfonated hydrocarbon, the mixture becomes dimly cloudy (colloidal suspension), and the cloud intensifies during the next 30 to 50 seconds. Finally, the insoluble colloidal AEA salt begins to coagulate and precipitate.

Once the insoluble calcium salt of the AEA ceases to be colloidal in nature and starts to coagulate and form curds, the ability of the admixture to entrain air is considerably reduced. An example of this phenomenon is illustrated in Table 6-2. The author has concluded that the difference in the rate of formation of the colloidal particles is essentially why the resin and rosin based AEAs perform better when the concrete is subjected to a short mixing period prior to placement, such as in concrete paving operations.

Table 6-2.   Effect of the Nature of the Calcium-Anionic AEA Salt on Its Ability to Entrain Air in Concrete.

CONCRETE NO.[a]	AEA[b]	W/C	SLUMP IN.	AIR %
1	Added directly to concrete with mix water	0.42	3-3/4	5.8
2	Allowed to stand 2 minutes in saturated lime water prior to addition with mix water	0.42	2.0	1.9

[a]Type I, C.F. = 517 lb
[b]Sulfonated hydrocarbon, 2 fluid oz/100 lbs cement

FACTORS THAT INFLUENCE AIR ENTRAINMENT

Of all the many things that have been learned about concrete through the proper use of admixtures, the entrainment of air is the most perplexing. The reason for this lies in the many factors that influence the amount of air that can be entrained by a given quantity of a given AEA. Some of these factors and their effects are summarized in Table 6-3. While these variables have been derived from the author's many

Table 6-3.   Factors That Affect Air Entrainment.

FACTOR	RESULT
Cement	• An increase in the fineness of cement will decrease the air content. • A high C.F. concrete will entrain less air than a lean mix. • An increase in the alkali content of cement will increase air entrainment.
Fine Aggregate	• Well rounded particles are conducive to air entrainment. • An increase in the fine fraction (passing the No. 100 sieve) will decrease the amount of entrained air. • An increase in the middle fractions (passing No. 16 but retained on No. 30 and No. 100 sieves) will increase the air content.
Coarse Aggregate	• Dust on the aggregate will lower the air content. • Crushed stone concrete will entrain less air than a gravel concrete.
Water	• Small quantities of household or industrial detergents in the water will increase the amount of entrained air. • If hard water (well or quarry) is used to dilute the AEA prior to batching, the air content will be reduced.

Table 6-3. (continued).

FACTOR	RESULT
Slump	• An increase in slump from 3″ to about 6″ will increase the air content. Above this, the air becomes less stable and the air content drops. A decrease in slump beyond 3″ makes air more difficult to entrain.
Temperature	• An increase in concrete temperature will decrease the air content. Increases from 70° F to 100° F may reduce air contents by 25%, while reductions from 70° F to 40° F can increase air contents by as much as 40%.
Concrete Mixer	• The amount of air entrained by a given mixer (stationary, paving, transit) will decrease as the mixing blades become worn. • Air content will increase if the mixer is loaded to less than capacity. However, in very small loads, such as a laboratory drum mixer, air becomes more difficult to entrain.
Mixing Time	• The air content will increase with increased time of mixing up to about 2 minutes in stationary or paving mixers and to about 15 minutes in most transit mixers.
Vibration	• Excess vibration will reduce air contents. As much as 50% of entrained air may be lost after 3 minutes of vibration.
Pozzolans	• As the fineness of the pozzolan increases the amount of entrained air decreases. • As the carbon content of the pozzolans increases, the amount of entrained air decreases. • An increase in the amount of pozzolanic material/yd^3 of concrete will decrease the amount of entrained air.
Oil and/or Grease	• Depending on their composition, they will either increase or decrease the amount of entrained air. These organic impurities usually occur in concrete as a result of poor lubricating practices at the cement plant or ready mix plant.
Chemical Admixtures	• Most chemical admixtures will increase the amount of entrained air when added in conjunction with an AEA. Type C admixtures usually do not affect the air content, although the addition of straight calcium chloride may tend to reduce the amount of entrained air. • Delaying the addition of either the chemical admixture or the AEA by as little as 15 seconds will increase the amount of air entrained.

years of experience, an excellent summary of these and more have been reported elsewhere [17][18].

Some of the factors listed in Table 6-3 warrant further discussion. First, the effect of the fineness of the portland cement on the amount of air incorported in concrete, both plain as well as that treated with an AEA, is shown in Table 6-4. A number of theories have been proposed to account for the fineness effect on air entrainment, but the author favors the concept that says that the fines tend to penetrate or puncture the protective skin around the entrained air void and reduce the film's effectiveness in protecting the air voids from coalescing. Dusts of high fineness, such as clay and limestone, introduced by the coarse aggregate, extremely fine particles in sand and those in pozzolanic materials would have a similar effect.

Next, consider the role of the increased alkali content of concrete, whether it be introduced via the cement, pozzolan or other ingredients, on the concrete's response to air entrainment. Alkalies tend to depress the solubility of calcium ions in the aqueous phase of concrete [19]. The calcium-anionic AEA films that surround the air void in the freshly mixed concrete are probably thinner (or possibly more soluble) than those which would develop in concretes of low alkali content. While the thinner film produced per unit AEA added permits more air to be entrained during the initial mixing, it also reduces the stability of the air entrained air void because the thinner film has a lower resistance to the pressure of the air within the bubble. As a result, as the mixing time is extended, such as in its transportation to the site of placement, the amount of entrained air decreases.

Concretes were fabricated from 11 different portland cements hav-

Table 6-4. Effect of Fineness of Portland Cement on the Air Content of Concrete.

	AIR CONTENT OF CONCRETE[b]	
CEMENT FINENESS-CM^2/G[a]	PLAIN	AEA ADDED[c]
3670 (as received)	1.9	5.9
4540 (ground)	1.2	4.3
6010 (ground)	0.8	3.8

[a]The same cement was used in the three concretes, but ground to the higher fineness values in laboratory steel ball mills. Fineness-method of Blaine.
[b]Concretes had a C.F. = 517 lb and a slump of 3-1/2 ± 1/4″.
[c]AEA addition rate = 2 fluid oz/100 lb AEA based on a sodium salt of sulfonated hydrocarbon.

ing various alkali contents but essentially the same fineness (3200 to 3400 cm^2/g). All of the concretes had a C.F. of 634 lb, were batched to a 3-1/2 ±1/2 in. slump, and treated with 1 fl oz of anionic AEA per 100 lb of cement, added with the gage water. The air content of each concrete was measured after the prescribed ASTM laboratory concrete mixing cycle. The amount of soluble alkali contributed by each cement to the aqueous phase of the concretes was then determined after rapidly mixing each cement with water, at the same water-cement ratio as that employed in preparing each of the respective concretes—for 15 seconds and for 8 minutes. The results of that work are shown in Figures 6-3A and 6-3B. The best straight line through each set of data points was determined by computer methods, and the slopes of those straight lines are 0.83 (Figure 6-3A) and 0.93 (Figure 6-3B). While the data points are scattered, the trend, i.e., an increase in alkali (soluble) by the cement results in an increase in air entrainment, is there. A great deal of this scatter is attributed to the analytical techniques some 25 years ago.

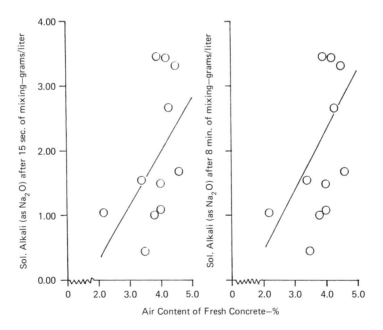

Figure 6-3A and 6-3B. Effect of cement alkali (soluble) on air entrained in concrete.

FOAM INDEX TEST

The author described a fairly simple test designed to evaluate the influence of *cementitious* materials on the air entraining capacity of concrete containing those substances at an informal meeting of the Transportation Research Board, in Washington, D.C. in January, 1980. The test became known as the Foam Index Test and has undergone a number of modifications since that time by himself and others [20][21][22]. Although the test was originally thought of as being a quality control procedure for pozzolans (fly ash, silica fume, etc.), it has since been found to be valuable in detecting the source of air entraining problems in concrete in which pozzolans are absent.

The Foam Index Test has four procedural streps:

1. Dilute any commercially available AEA 1:20 with water (1 part AEA, 20 parts distilled water, by volume) and retain the solution as a standard for subsequent tests.
2. Place a 20.0 gram sample of the cement to be tested in a clean, dry 4 fluid oz glass bottle. Add 50 mL of distilled water to the bottle, cap the bottle and shake its contents for 1 minute to thoroughly wet the cement.
3. Add the standard diluted AEA to the bottle in 0.2 mL aliquotes from a burette. After each addition, cap the bottle and shake it vigorously for 15 seconds. Then lay the bottle on its side (or upright) and observe the stability of the foam in the bottle.
4. The amount of the dilute AEA, in mL, needed to produce a stable foam, i.e., one in which no bubbles can be seen breaking for a period of 15 seconds, is the foam index of the cement.

The photograph in Figure 6-4 illustrates the Foam Index Test in progress. The bottle on the left has only been treated with a few volume increments of the diluted AEA and the foam initially formed after shaking had disappeared less than 5 seconds after the bottle had been placed on its side. The residue on the inner surface of the bottle, above the liquid level, is the result of the bubbles in the unstable foam bursting and depositing their adhering solids on the inner walls of the bottle. The bottle on the right in Figure 6-4 is typical of the end-point of the test. The foam is stable and rides high and thick on the surface of the liquid phase, and there is no visible collapse of the bubbles

Figure 6-4. Foam Index Test in progress.

during the 15 second observation period. The end-point is sharp, but the test does require patience on the part of the operator.

In order to use the Foam Index Test as a means of evaluating the effect of portland cement on air entrainment in concrete, one must first have a reference point. The author chose a cement that performed adequately in air entrained concrete as the standard, i.e., one having a fineness of 3550 cm^2/g and a total alkali content of 0.65% (as NA_2O). In a 517 lb C.F. concrete treated with 1 fluid oz/100 lbs cement of the AEA (also used in the Foam Index Test), the amount of air in the fresh concrete was found to be 5.4% and its slump was 3-1/4 inches. The Foam Index of the cement was found to be 9.4 mL. That same cement was then ground in laboratory steel ball mills to three additional degrees of fineness and the products of those grinds were evaluated in concrete and by the Foam Index Test. The results are summarized in Table 6-5 and those relationships are illustrated in Figure 6-5. The results are in agreement with those previously shown in Table 6-4 which vividly indicate that an increase in the fineness of cement lowers its response to air entrainment. Note also that as the fineness of the cement increases, its Foam Index increases.

Four cements having essentially the same fineness (3300–3400 $cm^2/$g) but having a wide range of total alkali contents were evaluated in concrete using a fixed addition rate of an AEA. The foam index

Table 6-5. Cement Fineness, Foam Index, and Air Contents of Concrete.

FINENESS OF CEMENT-CM^2/G	AIR CONTENT FRESH CONCRETE[a]	FOAM INDEX OF CEMENT-ML
3550	5.4	9.4
3910	5.1	10.0
4260	4.8	10.4
5020	4.0	11.0

[a]Addition rate of AEA = 1 fluid oz/100 lb of cement.
Slump of all concretes = 3-1/2 ± 1/4"

of each cement was also determined using a 1:20 dilution of the same AEA (Table 6-6 and Figure 6-6). The data scatter in Figure 6-6 is essentially due to the assumption that all of the alkali in the cement was readily soluble in the aqueous phase of the concrete and in the cement-water mixture in the Foam Index Test. However, the data do show that as the alkali content of cement increases, its ability to entrain air in concrete increases (previously shown in Figures 6-3A and 6-3B) and its foam index decreases.

The presence of contaminants in cement that disturb its normal response to air entrainment in concrete can also be detected by the measurement of its foam index. For example, a cement whose fineness was 3710 cm^2/g and total alkali content was 0.76% (as Na$_2$O) exhibited poor response to air entrainment in concrete, i.e. 1 fluid oz/100 lbs of cement produced only 2.9% air in the fresh concrete. It was expected that the amount of air would be in the range of 5.0%–

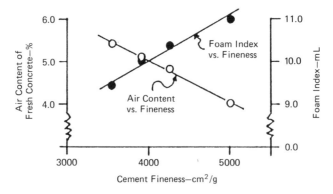

Figure 6-5. Effect of cement fineness on air content of fresh concrete and Foam Index.

Table 6-6. Effect of Cement Alkali Content and Foam Index on Air Content of Concrete.

ALKALI CONTENT OF CEMENT-% (AS NA₂O)	AIR CONTENTS OF FRESH CONCRETE-%[a]	FOAM INDEX OF CEMENT-ML
0.32	4.5	9.6
0.68	5.3	8.0
0.72	6.1	7.2
0.93	6.8	6.6

[a]Addition rate of AEA = 1 fluid oz/100 lb of cement.
Slump of all concrete = 3-1/2 ± 1/4".

5.5%, knowing the cement's fineness and alkali content. The foam index of the cement was 15.4 mL and should have been in the range of 8.4–9.4 mL. The cement was then subjected to an ether-acetone extraction and found to contain a small amount of oil (0.003%), and this was attributed to its poor response to air entrainment. Attempts to carry this concept one step further to cover coarse and fine aggregates have produced mixed results.

STABILITY OF ENTRAINED AIR

It was mentioned earlier that an increase in the alkali content of portland cement had a definite influence on the stability of the entrained air void. A typical example of this is illustrated by the data in Table 6-7. While the larger decrease in the slump of concrete No. 2 over

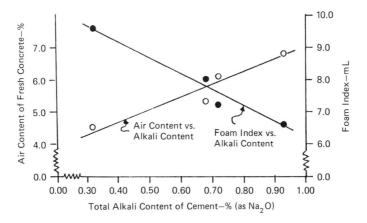

Figure 6-6. Influence of alkalie on air entrainment and Foam Index.

Table 6-7. Effect of Alkali on the Stability of Entrained Air.

CONCRETE NO.[a]	SOLUBLE ALKALI IN CEMENT (AS NA₂O)-%[b]	AMOUNT OF ENTRAINED AIR—%[c]		
		INITIAL	AFTER 30 MIN.	DIFFERENCE
1	0.32	5.2	4.8	0.4
2	0.82	7.5	4.1	3.4

[a]C.F. = 517 lb, initial slump = 4″.
[b]After 8 minutes of rapid mixing with water-cement ratio equal to that of the respective concrete. The two cements had the same fineness (3310 cm^2g).
[c]AEA, a sulfonated hydrocarbon 2 fluid oz/100 lb cement.

that of No. 1 (by approximately 1-1/2″) after the 30 minute mixing period could have contributed to its reduction in air content, the increased alkali is probably the principal culprit in this phenomenon, in one way or another.

EFFECT OF ENTRAINED AIR ON THE PROPERTIES OF FRESH CONCRETE

The presence of entrained air in fresh concrete has a pronounced effect on its properties. One of these is workability, which is improved. For adequate workability, the aggregate particles must be spaced in a way that they can move past one another with relative ease during mixing and placing. In this respect, the entrained air voids are often thought of as acting as millions of tiny ball bearings in the concrete, making the mix more workable. Many users of AEAs take advantage of this improved workability by reducing the water-cement ratio in the treated concrete to maintain the same workability as the plain, untreated concrete. An example of this effect is illustrated in Table 6-8. The author has found that for every 1% air entrained, the water-cement ratio can be reduced by approximately 1% *up* to an amount of air that is in the range of 5%–6% and still maintain the same workability. In cement rich concretes (C.F. = 600 + lbs) this 1:1 ratio drops to a lower value, while in leaner mixes (C.F. = 500 − lb) that ratio increases.

The presence of entrained air also affects the bleeding of the concrete. Bleeding is the autogenous flow of mixing water within, or emergence from, newly placed concrete (or mortar) caused by the settlement of solid materials within the mass [23]. Bleeding is basically the result of segregation, i.e., separation of coarse aggregate from the mortar or separation of the paste from aggregate, which re-

Table 6-8. Effect of Air Entrainment on the Water-Cement Ratio and Workability (Slump) of Concrete.

CONCRETE NO.[a]	WATER-CEMENT RATIO	SLUMP-IN.	AIR CONTENT %[b]
1	0.54	3-1/2	1.7
2	0.53	3-1/4	3.5
3	0.52	3-1/2	5.6

[a]C.F. = 517 lb, Type 1 cement, 1/4″ maximum size coarse aggregate.
[b]Anionic AEA based on a sulfonated hydrocarbon.

sults in the appearance of a layer of water on the surface of the cementitious mass, after placement. Entrained air eliminates or minimizes segregation and subsequent bleeding by at least two mechanisms: the air bubbles (1) provide a certain degree of buoyancy to the aggregates and cement, reducing their rate of sedimentation and (2) reduce the effective volume through which the differential movement of water may occur.

Bleeding of fresh concrete (and/or mortar) can adversely affect certain properties of the hardened mass. First, bleeding indicates an increase in heterogeneity or non-uniformity and results in a lowered resistance to stress (compressive, flexural and tensile). Second, if the bleed water residing on the surface is not removed, the very high water-cement ratio on the surface is such that the surface will exhibit lower than normal strength after hardening and have a reduced resistance to wear and abrasion. On the other side of the coin, if the surface dries out due to an inadequate amount of water, the early hydration of the cement is impeded and this leads to the same results. So, there is a delicate balance between too much water and not enough water on the surface. This is why moist curing, fog mist, or saturated burlap after removal of the bleed water is recommended for all concretes.

EFFECT OF ENTRAINED AIR ON THE PROPERTIES OF HARDENED CONCRETE

Another misconception about air entrainment has to do with the *amount* of air entrained in concrete. For example, many users are prone to think that if a little air in concrete is good, a lot of air is better. *Wrong!* Up to 5%–6% entrained air in concrete is good, but beyond that range, the amount of entrained air can be deleterious to the structural properties of the hardened concrete. And, at this point the author would like to present his views on the positive as well as negative effects on those properties.

The relationship between the amount of air measured in fresh concrete and that determined in the hardened mass, by microscopical methods, should first be considered [24]. The difference in the amount of air in 58 freshly mixed concretes and that determined in the respective hardened concrete averaged 1.3% (lower). It should be pointed out that the value measured in the fresh concrete was its total air while only those air voids having a chord length of 1 mm or less were considered entrained. In only one case did the air entrained in the hardened concrete exceed the total air content measured in the fresh concrete. This situation is attributed to a gross error made during either or both measurements. Those 58 test concretes were fabricated from various types and brands of cement, several different anionic AEAs and at C.F.s ranging from 517–635 pounds.

The relationships between the air content measured in the fresh concrete (type I cement, C.F. = 517 lbs) and those found in the hardened concrete (1 mm or less in chord length) for eight mixes are shown in Figure 6-7. Four of the concretes were treated with one type of anionic AEA and the second four with another. It is quite evident from the data in Figure 6-7 that more of the total air in the fresh concrete containing Brand A AEA appears as entrained air in the hardened concrete than that when Brand B AEA is used. That is one of the reasons why the author has chosen to take the difference (total air, fresh concrete-air entrained in hardened concrete) value as being 1.3%, cited earlier, but remembering that this is only an average of less than one hundred determinations.

The results of the author's work in the early 1970s brought to light a very important point. The amount of entrained air in hardened laboratory concrete has been (and still is) determined from laboratory cast and cured 4 × 8″ cylinders. The author found that in order to obtain a true picture of the entrained air content in the test specimen, it should be cut its entire vertical length before polishing and subsequent microscopic examination. The amount of entrained air in the bottom, middle and top third of a typical air entrained concrete specimen is shown in Figure 6-8. Up until that time it was assumed that

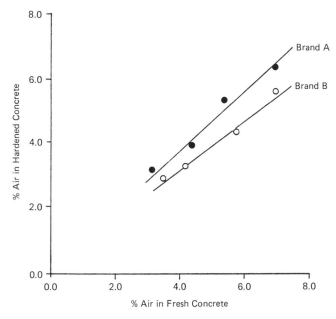

Figure 6-7. Relationship between the measured air contents of fresh concretes and those of the respective hardened concretes, air entrained by two brands of anionic AEAs.

the horizontal portion cut from the middle of the concrete cylinder could be used as a guide to its entrained air content. In this case, the 5.6% entrained air measured in that portion of the specimen was very close to the average value of 5.8% cited in Figure 6-8.

If one takes advantage of the reduction in water-cement ratio pro-

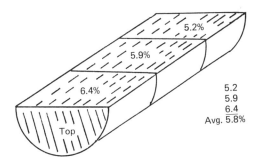

Figure 6-8. Air content of a hardened 4″ × 8″ concrete cylinder.

vided by air entrainment, that reduction will more than compensate for the amount of entrained air in the hardened concrete, up to 5%– 6% air. This is especially true for lean mix concretes. Once that range is exceeded, the reduction in water-cement ratio cannot compensate for the number of discontinuities (or voids) within the paste, and the compressive strength of the concrete drops rather sharply. A typical example of this is illustrated in Figure 6-9. Another cause for this decrease in strength is the tendency of the air voids to collect at the surfaces of the large aggregate when their number exceeds a certain range which reduces the strength of the paste to aggregate bond. Microphotographs of two concretes, one containing properly entrained air and one in which a large number of the air voids are located at the paste-aggregate interface are shown in Figures 6-10 and 6-11, respectively. All of the concretes in Figure 6-9 were batched to a 3-1/2″ ± 1/4″ slump, contained 3/4″ top size aggregate, had a C.F. of 430 lb and were cured under standard laboratory conditions.

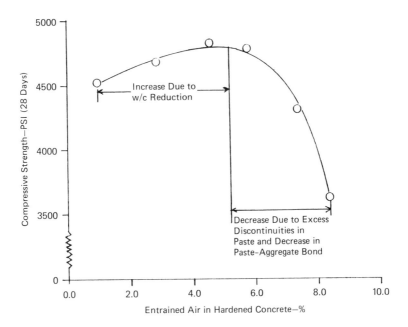

Figure 6-9. Effect of entrained air content on the 28-day compressive strength of concrete.

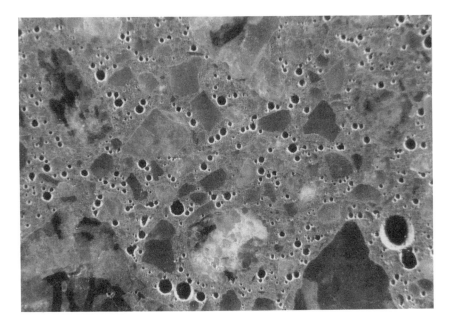

Figure 6-10. Microphotograph of concrete containing properly entrained air (5.2% air, 10X).

CONCRETE FREEZE-THAW DURABILITY

Although there are other advantages to be realized, the principal reason behind the use of AEAs is the improvement that they impart to the concrete's resistance to frost action and to its deterioration due to repeated freezing and thawing. For the sake of this discussion the author will consider the two as being the essentially same and will lump them together as freeze-thaw (F-T) durability. Although surface scaling might be included in this "lump" other factors such as premature finishing, play an important role in that particular phenomenon.

Contrary to common thinking, structural concrete is quite porous, in spite of its specific gravity which can range from 145–165 lbs/ft^3, after hardening. There are basically two kinds of pores, capillary and gel, within the hardened mass. The capillary pores are approximately 0.5 microns (μm), or 2×10^{-5} inches in diameter while the diameter of the gel pores are close to 30 Angstroms (Å) or 13×10^{-8} inches.[a]

---

[a]1 micron μm $= 1 \times 10^{-4}$ cm $= 4 \times 10^{-5}$ inch
1 Angstrom Å $= 1 \times 10^{-8}$ cm $= 4 \times 10^{-9}$ inch

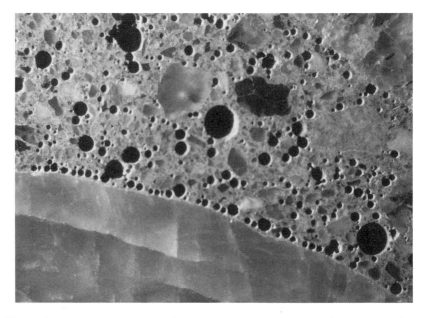

Figure 6-11. Microphotograph of concrete containing an excessive amount of entrained air (7.5% air, 16X).

According to the modern theory of nucleation, crystals cannot grow in a melt unless the nuclei exceed a certain critical size. Since the dimensions of the gel pores are less than the critical size for ice nuclei, water in these pores will not freeze, even at temperatures as low as −100° F. The diameter of the capillary pores, however, does exceed the critical size for ice nucleation, and water in such pores will freeze when the temperature of the concrete drops to 32° F or below.

In a saturated hardened paste, both the gel and capillary pores are filled with water. When the temperature of the concrete decreases to a point at which freezing can occur, ice crystals form in the capillary pores. As water freezes, expansion occurs, and its increase in volume, over the liquid water is approximately 10%. Water is one of the few substances exhibiting this unusual behavior of expansion on freezing and will cause either dilation of the capillary pore or expulsion of water from it. If there is an empty air void close to the capillary pore, the excess water can escape by way of the gel pores into that void. The growing crystals of ice in the capillary pores act as pumps that force the water through the gel pores towards the nearest air void boundary, thus creating a hydraulic pressure in the system. The pressure controlling factor is the distance from the capillary pores to the air void boundaries. If the distance is too great, the pressure can become so great that it dilates the gel pores and ruptures the structure of the paste. The concrete then exhibits a gradual loss of integrity and strength and ultimately its deterioration.

It is generally accepted by the concrete industry that in order for concrete to withstand the rigors of freezing and thawing, it must contain entrained air voids, and the *spacing* factor of those voids must be 0.0080 inches or less. The *spacing factor* is defined as the maximum distance of any point in the paste or in the cement paste fraction of mortar or concrete from the periphery of an air void [25]. The spacing factor is calculated from the measured (or calculated) paste content and air void content of the concrete, and the average specific surface of the air voids as shown in Equations 6-3 and 6-4 [8].

$$L = p / \ 400 \ n \ \text{when } p/A \text{ is less than or equal to } 4.342 \qquad (6\text{-}3)$$

$$L = \frac{3}{\alpha} \left[1.4 \left(\frac{p}{A} + 1 \right)^{1/3} - 1 \right] \qquad (6\text{-}4)$$

when p/A is greater than 4.342
    where
    L = spacing factor
    p = paste content, volume percent of concrete.
    n = average number of air voids intersected per linear inch (or millimeter) of traverse.
    α = specific surface of air void = 4/e
    e = average chord length of air voids in inches (or millimeters) traversed and equal to A/100n
    A = air void content

The spacing factor of the entrained air voids is only one of two major factors that play an important role in determining F-T durability. The other is its strength at the time that it is first subjected to the F-T environment. Many concretes placed after October 1st, and even as early as September 15th of a given year in the northern regions of the U.S.A. frequently exhibit F-T deterioration in the Spring that follows, although their air contents and air void distributions would have been expected to be adequate to guard against deterioration. The author attributes this to the persistently lower temperatures of the environment, both during the day and night time hours, preventing the concrete from attaining sufficient strength to ward off the F-T forces that were to develop during the following winter months.

The author's studies have shown that the F-T durability of concrete is directly related to its compressive strength, at the time that it is subjected to the hostile environment and inversely related to the spacing factor of the entrained air voids.

$$\text{F-T D.I.} = \frac{\text{Compressive strength at time of initial exposure to test} - \text{PSI}}{\text{Spacing factor in.}} \times K$$

(6-5)

That relationship is expressed in Equation 6-5 and is called the F-T Durability Index (F-T D.I.) of concrete. The author arbitrarily chose the value of K, a constant, to be 1000 so that F-T D.I. would be a reasonably small number. The F-T D.I. is an indication of whether a given air entrained concrete will pass or fail the test for durability [26].

In one experiment, 22 concretes were fabricated from four (4) different cements at various C.F.s and having different entrained air contents, produced by two (2) different anionic AEAs. Each concrete was cast into the standard F-T beams and into 4 × 8″ cylinders. The beams and cylinders were moist cured in their molds for 24 hours and after de-molding all the test specimens (beams and molds) were immersed in saturated lime water for a period of 13 days (72° F). At the end of the 14-day curing cycle, one cylinder from each concrete batch was cut longitudinally and subjected to hardened air analysis by the method of linear traverse. The remaining cylinders from each of the test concretes were then broken in compression just before the beams were subjected to the F-T test (ASTM C666, Method A, freezing and thawing in water). The author assumed that the compressive strengths of the test cylinders were representative of the concrete in the beams and that during the F-T cycling, between 0° F and 40° F, the strength of the concrete in the beams would undergo very little change.

A description of the 22 concretes is shown in Table 6-9, and a plot

Table 6-9.   F-T Durability of Air Entrained Concrete.

CONCRETE NO.	C.F.- LB/YD^3	HARDENED AIR-%[e]	SPACING FACTOR -IN.	RDME[f] (P)-%	COMPRESSIVE STRENGTH PSI	F-T D.I.	PASS (P)[g] OR FAIL (F)
1	545[a]	4.9	0.0081	95	4350	537	P
2	530[a]	5.8	0.0068	96	4200	618	P
3	510[a]	5.5	0.0072	89	3960	550	P
4	475[a]	6.0	0.0068	80	3810	560	P
5	425[a]	5.2	0.0075	54	3400	453	F
6	525[b]	5.0	0.0078	95	4530	581	P
7	490[b]	5.9	0.0071	90	4010	565	P
8	455[b]	4.9	0.0078	75	3550	455	P
9	410[b]	5.5	0.0076	56	2910	383	F
10	380[b]	4.7	0.0085	50	2730	321	F
11	517[c]	5.9	0.0072	82	3770	524	P
12	465[c]	4.7	0.0079	68	3720	471	P
13	420[c]	4.5	0.0079	57	2910	368	F
14	380[c]	5.0	0.0070	55	2610	373	F
15	355[c]	5.4	0.0074	50	2150	290	F
16	550[d]	5.8	0.0068	99	4950	728	P
17	513[d]	6.2	0.0060	98	4100	683	P
18	485[d]	5.0	0.0081	68	3840	474	P
19	455[d]	5.6	0.0075	58	3410	455	F
20	412[d]	5.0	0.0070	56	2880	411	F
21	500[c]	10.6	0.0048	57	2330	485	F
22	450[c]	10.9	0.0046	50	2000	435	F

[a]Type I cement—Midwest USA
[b]Type I cement—Southwest USA
[c]Type II cement—Northeast USA
[d]Type III cement—Northeast USA
[e]Chord length of air void—1 mm or less
[f]$P = n_1^2/n_2^2 \times 100$
n_1 = fundamental transverse frequency at 0 F-T cycles
n_2 = fundamental transverse frequency after 300 F-T cycles
[g]P must be greater than 60% at the end of 300 F-T cycles in order for the concrete to pass the ASTM C666 test.

of the concretes' F-T D.I., as a function of their relative dynamic modules of elasticity at the end of 300 F-T cycles is illustrated in Figure 6-12. If the minimum F-T D.I. is set at a value of 450 as a guide as to whether a given air entrained concrete will meet the ASTM C666 requirements, only 13.6% (3 out of 22) of the data points fail to correlate with the chosen lower limit. The value of 450 for the F-T D.I. seems to represent a reasonable cut-off point (pass or fail), considering (1) the relatively small number of test concretes and (2)

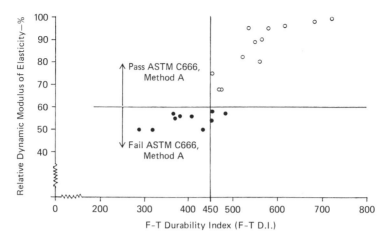

Figure 6-12. Effect of F-T D.I. on the response of air entrained concrete to freezing and thawing cycles.

the precision of the test procedures and methods involved; i.e., measurement of compressive strength and modulus of elasticity, the eccentricities of the F-T test and finally the linear traverse analysis. The author sincerely hopes that any reader who has performed similar work will come forward via publication or correspondence to either refute or substantiate the concept just presented.

Before leaving Table 6-9, it should be pointed out that Concretes No. 21 and 22 were designed to prove the point that a little air is good for concrete but a lot of air can be detrimental. It is possible to entrain so much air in concrete that its compressive strength at the time of F-T testing is so low that its F-T D.I. is less than 450 resulting in failure of the F-T test.

ECONOMICS OF USING AIR ENTRAINING ADMIXTURES

The cost of entraining air in concrete is minimal or it may even reduce the overall cost of the concrete, depending upon the costs of the raw materials and labor, which vary from one geographic area to another, and the concrete design philosophy of the concrete producer and/or specifying agency. It is of interest to make a series of cost calculations based on the arbitrarily chosen assumptions listed in Table 6-10.

Table 6-10. Assumptions Made in Calculating Cost of Concrete.

RAW MATERIAL	COST $	SPECIFIC GRAVITY
Cement	0.040/lb	3.15
Aggregage	0.005/lb	2.52
AEA	0.010/fluid oz	1.00
Water	—[a]	

[a]The cost of water has been considered negligible.

First, as a base for comparison, consider the cost of plain, untreated concrete, per cubic yard having a slump of 4 $\pm 1/2$ inches (concrete No. 1).

Mix Design (No. 1)

Cement 500 lb/yd^3 = 2.544 ft^3/yd^3
Water 275 lb/yd^3 = 4.407 ft^3/yd^3
(water-cement ratio = 0.55)
Air 1.5% (no AEA) = 0.405 ft^3/yd^3
Aggregate 3090 lb/yd^3 = 19.644 ft^3/yd^3
(wt. of coarse:fine agg = 55:54)
Yield = 27.000 ft^3/yd^3

Raw Material Cost

$35.45/yd^3 (from costs in Table 6-10)

Concrete No. 2 is an air entrained version of No. 1 but has been designed to take advantage of the water reduction that will accrue and thus off-set any reduction in strength caused by the entrained air. In other words, concrete No. 2 is to have the same slump as concrete No. 1.

Mix Design (No. 2)

Cement 500 lb/yd^3 = 2.544 ft^3/yd^3
Water 265 lb/yd^3 = 4.247 ft^3/yd^3
(water-cement ratio = 0.53)
Air 6% (1% entrapped, 5% entrained) = 1.620 ft^3/yd^3
Aggregate 2925 lb/yd^3 = 18.589 ft^3/yd^3
AEA—1-1/4 fluid oz/100 lb of cement
Yield = 27.000 ft^3/yd^3

Raw Material Cost

$34.69/yd^3 (from costs in Table 6-10)
Cost Savings = $35.45 (No. 1)—$34.69 (No. 2) = $0.76/yd^3
The amount of air entrained in the paste is

$$\frac{1.350 \text{ ft}^3/\text{yd}^3 \text{ (entrained air)}}{8.411 \text{ ft}^3/\text{yd}^3 \text{ (total air + water + cement)}} \times 100 = 16.1\%$$

Concrete No. 3, also air entrained, is a redesign of No. 1 to take advantage of the increase in workability afforded by the entrained air; i.e., the same water-cement ratio as that of concrete No. 1. Since the entrained air under these conditions will lower the strength of the concrete, the amount of cement will have to be increased.

Mix Design (No. 3)

Cement 530 lb/yd^3 = 2.696 ft^3/yd^3
Water 291 lb/yd^3 = 4.671 ft^3/yd^3
(water-cement ratio = 0.55)
Air 6.5% (1% entrapped, 5.5% entrained) = 1.755 ft^3/yd^3
Aggregate 2810 lb/yd^3 = 17.878 ft^3/yd^3
AEA − 1-1/4 fluid oz/100 lb of cement
Yield = 27.000 ft^3/yd^3

Raw Material Cost:

$35.32/yd^3 (from costs in Table 6-10)
Cost Savings = $35.45 (No. 1) − $35.32 (No. 3) = $0.13/yd^3
More air was added to concrete 3 than to No. 2 in order to attain essentially the same volume percent air in the paste as in No. 2.

$$\frac{1.485 \text{ ft}^3/\text{yd}^3 \text{ (entrained air)}}{9.122 \text{ ft}^3/\text{yd}^3 \text{ (total air + water + cement)}} \times 100 = 16.3\%$$

Although it is easy for any one to design a concrete mix on paper, only trial mixes will establish the actual amount of raw material needed to attain a specified strength, slump and air content. For example, the reduction in water-cement ratio in No. 2 might not be sufficient to maintain the slump requirement and compensate for the influence of the air on the strength. So, additional cement will be needed and the raw material cost of the concrete will increase and possibly exceed that of No. 1. Although the cement content in No. 3 was increased by only 30 lb/yd^3, additional cement may be needed. Even if the raw material cost of concretes No. 2 and 3 exceed that of No. 1 by as much as $0.25/yd^3, it is a small price to pay for durability and minimizing the detrimental effects of bleeding. The author views the cost of using AEAs as being a very low cost form of insurance against the horrendous cost of the three R's; repair, removal and replacement.

REFERENCES

[1] Powers, T. C., "The Bleeding of Portland Cement Paste, Mortar and Concrete," *Research Department Bulletin No. 2,* Portland Cement Association, pg. 97 (1939).

[2] Swayze, M. A., "More Durable Concrete with Treated Cement," *Engineering News-Record,* Vol. 126, pp. 946–949 (1941).

[3] Anderson, A. A., "Experimental Test Data in Concrete with the Development of Chloride Resisting Concrete by the Use of Treated Portland Cements and Blends with Natural Cement," *17th Annual Proceedings,* Association of Highway Officials of the North Atlantic States, pp. 67–88 (1941).

[4] ASTM C138, "Standard Test Method for Unit Weight, Yield and Air Content (Gravimetric) of Concrete," *Annual Book of ASTM Standards,* Vol. 04.02, pp. 80–82 (1988).

[5] ASTM C173, "Standard Test Methods for Air Content of Freshly Mixed Concrete by the Volumetric Method," *Annual Book of ASTM Standards,* Vol. 04.02, pp. 106–109 (1988).

[6] ASTM C231, "Standard Test Method for Air Content of Freshly Mixed Concrete by the Pressure Method," *Annual Book of ASTM Standards,* Vol. 04.02, pp. 131–137 (1988).

[7] Brown, L. S., Pierson, C. U., "Linear Traverse Technique for Measurement of Air in Hardened Concrete," *Proceedings,* American Concrete Institute, PACIA, Vol. 47, pp. 117–123 (1950).

[8] ASTM C457, "Standard Practice for Microscopical Determination of Air-Void Content and Parameters of the Air-Void System in Hardened Concrete," *Annual Book of ASTM Standards,* Vol. 04.02, pp. 222–236 (1988).

[9] Powers, T. C., "The Air Requirement of Frost-Resistant Concrete," *Proceedings,* Highway Research Board, Vol. 29, pp. 184–202 (1949).

[10] Powers, T. C., "Void Spacing as a Basis for Producing Air-Entrained Concrete," *ACI Journal, Proceedings,* Vol. 50, No. 9, pp. 741–760 (1954).

[11] ASTM C260, "Standard Specification for Air-Entraining Admixtures for Concrete," *Annual Book of ASTM Standards,* Vol. 04.02, pp. 152–154 (1988).

[12] ASTM C233, "Standard Test Method for Air Entraining Admixtures for Concrete," *Annual Book of ASTM Standards,* Vol. 04.02, pp. 143–146 (1988).

[13] Gutman, P. F., "Bubble Characteristics as They Pertain to Compressive Strength and Freeze-Thaw Durability," *ACI Materials Journal*, pp. 361–366 (1988).

[14] "Cement and Concrete Terminology," *American Concrete Institute*, Publication SF-19, pg. 8 (1988).

[15] Mielenz, R. C., Wolkodoff, V. E., Backstrom, J. E., Flack, H. L., "Origin, Evaluation and Effects of the Air System in Concrete. Part 1-Entrained Air in Unhardened Concrete," *Journal of the American Concrete Institute*, pg. 118, July (1958).

[16] ASTM C185, "Standard Test Method for Air Content of Hydraulic Cement Mortar," *Annual Book of ASTM Standards*, Vol. 04.01, pp. 181–184 (1986).

[17] Anonymous, "Control of Air Content," *Concrete Construction*, pp. 717–724, Aug. (1984).

[18] Gaynor, R. D., "Control of Air Content in Concrete—An Outline," *National Ready Mixed Concrete Association and National Sand and Gravel Association*, pp. 1–11 (1977).

[19] Hansen, W. C., Pressler, E. E., "Solubility of $Ca(OH)_2$ and $CaSO_4 \cdot 2H_2O$ in Dilute Alkali Solutions," *Industrial and Engineering Chemistry*, Vol. 39, No. 10, pp. 1280–1282 (1947).

[20] Meininger, R. C., "Use of Fly Ash in Concrete-Report of Recent NSGA-NRMCA Research Laboratory Studies," distributed at NRMCA Quality Control Conference, St. Louis MO, July (1980).

[21] "Improvement of the Foam Index Test," Project No. 7040-82-1, British Columbia Hydro and Power Authority Research and Development, March (1983).

[22] Private communication, R. D. Pavlovich, Engineers Testing Laboratory, Phoenix, AZ, to H. McGinnis, Western Ash Co., Fullerton, CA, July 24 (1980).

[23] "Cement and Concrete Terminology," *American Concrete Institute*, Publication SP-19, pg. 18 (1988).

[24] ASTM C457, "Standard Practice for Microscopical Determination of Air-Void Content and Parameters of the Air-Void System in Hardened Concrete," *Annual Book of ASTM Standards*, Vol. 04.02, pp. 222–232 (1988).

[25] "Cement and Concrete Terminology," *American Concrete Institute*, Publication SP-19, pg. 133 (1988).

[26] ASTM C666, "Standard Test Method for Resistance of Concrete to Rapid Freezing and Thawing," *Annual Book of ASTM Standards*, Vol. 04.02, pp. 406–418 (1988).

[27] Langan, B. W., Ward, M. A., "Determination of the Air-Void System Parameters in Hardened Concrete—An Error Analysis," *ACI Journal*, pp. 943–952, Nov.–Dec. (1986).

Chapter 7

POZZOLANS AND THE POZZOLANIC REACTION

INTRODUCTION

The most often used mineral admixture in the modern concrete industry is the pozzolan. A pozzolan, and there are many of them, is defined as "siliceous or siliceous and aluminous materials which in themselves possess little or no cementitious value but will, in finely divided form and in the presence of moisture, chemically react with calcium hydroxide at ordinary temperatures to form compounds possessing cementitious properties" [1]. This chemical reaction between the siliceous and/or siliceous-alumina components in the pozzolan, calcium hydroxide and water is called the *pozzolanic reaction.*

Two types of pozzolanic materials are readily available. There are the natural pozzolans which are of volcanic origin (and these were used by the early Romans and Greeks) such as trass, certain pumicites and perlite. Since the best of the many varieties of volcanic ash was found near Pozzoli, Italy, the material was called Pozzolana or Pozzolan (in English) and this name has since been extended to cover the entire class of mineral admixtures of which it is a member [2]. Those of the second type are man-made pozzolans which include such by-products as fly ash (the burning of coal), blast furnace slag (steel industry), and silica fume (silicon and ferrosilicon manufacture). Of the manmade pozzolans, fly ash is probably the most frequently used in concrete. Tracing the source of its name is, at best, difficult, but the author holds to the concept that (1) because it was ash, or a solid residue remaining when any combustible material is burned and (2) because it was allowed to fly out of the chimney, or smoke stack, with the hot flue gases (before the E.P.A. came into being), the name fly ash was adopted. And finally, fly ash has been defined by those of great wisdom as the "finely divided residue that results from the combustion of ground or powered coal" [3]. If the reader is a stickler on definitions, raw or calcined natural pozzolans include "such materials as some diatomaceous earths, opaline cherts, shales, tuffs, pumicites, any of which may or may not be processed by calcination and various materials requiring calcination to induce satisfactory properties, such as some clays and shales" [3]. In other words, naturally occurring pozzolans can be used as is or after processing (a reduction in particle size is often considered a part of the processing).

FLY ASH—CLASSES, COMPOSITION AND PROPERTIES

Fly ash will be considered first. Coal, from which it is derived, is made up of a mixture of elemental carbon, complex organic compounds and inorganic substances and when burned, energy is produced, the organic portion is converted to gases such as oxides of carbon and sulfur and water vapor. Since the inorganic portion is noncombustible, it is the principal component of the ash. The ash component of coal will range from 10% to 20% by weight of the coal. It should be pointed out here that not all of the ash derived from the burning of coal is fly ash. A good portion of it is what is often called "bottom ash." This is the ash that the author, as a youngster, had to remove daily from the basement of his parents' coal fired home. Bottom ash consists of large agglomerates, too large to be carried up the chimney with the flue gases.

Fly ash is described as being either Class C or Class F in ASTM C618 and the basic chemical requirements set forth for the two classes are listed in Table 7-1. The origin of the class designations C and F is not clear. The author speculates that Class C was assigned to those ashes which generally have a high calcium oxide (CaO) content and that Class F was chosen for those ashes which generally have a high iron oxide (Fe_2O_3) content. It was mentioned in Chapter 1 that the shorthand notation routinely used by cement chemists designates oxides by single letters, i.e., C = CaO and F = Fe_2O_3. Class F fly ash is derived from the burning of bituminous coal, while Class C fly ash is generated during the firing of subbituminous or lignitic coal. Almost all, if not all, of the coal mined east of the Mississippi River is of the bituminous type and very often a Class F fly ash is referred to as an Eastern ash. Since the coal mined west of that river is mainly

Table 7-1. Chemical Requirements of Fly Ash [3]

CHEMICAL COMPONENT	FLY ASH CLASS		
	C	F	N
$SiO_2 + Al_2O_3 + Fe_2O_3$—min %	50.0	70.0	70.0
SO_3—max %	5.0	5.0	4.0
Moisture content—max %	3.0	3.0	3.0
Loss on ignition—max %[a]	6.0	6.0	10.0
Available alkalies (as Na_2O)—max %[b]	1.5	1.5	1.5

[a]The use of a Class F fly ash containing up to 12.0% loss on ignition may be approved by the user if either acceptable performance records or laboratory tests are available.
[b]Optional chemical requirement.

subbituminous or lignitic, the Class C ash is frequently called a Western ash.

Typical compositions of a type I portland cement and of fly ashes representing Classes F and C are listed in Table 7-2. The purpose in presenting the data in Table 7-2 is three-fold. First, when either a Class F or C fly ash is added to portland cement concrete, the same kinds of metallic and non-metallic oxides as those of the cement are being added. In other words nothing new or strange is being introduced. Second, although the Class C ash appears to contain more calcium oxide (CaO) than the Class F, very little of it occurs as free, uncombined CaO, but as complex calcium silicates and aluminates which are not necessarily the same as those in portland cement, but somewhat akin to them; and, many possess some hydraulic properties. Returning to Table 7-2, it is important to note the high CaO

Table 7-2. Chemical Composition of a Typical Type I Portland Cement and of Fly Ash.

CHEMICAL COMPONENT	TYPE I CEMENT—%	CLASS F FLY ASH—%	CLASS C FLY ASH—%
SiO_2	19.8	43.4	32.5
Al_2O_3	6.1	18.5	21.9
Fe_2O_3	2.5	26.9	5.1
CaO	63.7	4.3	27.4
SO_3	2.2	1.2	2.8
MgO	3.5	0.9	4.8
Total Alkalies (as Na_2O)	0.9	0.6	1.1
Loss on ignition	1.0	3.2	1.2
Moisture	—	0.2	0.8

content of the typical type I cement and yet, the free, uncombined CaO in a portland cement seldom exceeds one percent. The Class C ash described in Table 7-2 contained only 2.9% free, uncombined calcium oxide [4]. Third, the loss on ignition (L.O.I.) of a Class C fly ash is, as a rule, lower than that of Class F ashes. Ninety-nine percent, or more, of the L.O.I. of a given fly ash is due to unburned carbon in the ash and its presence can be detrimental when concrete to which the ash has been added is treated with an air entraining admixture (Chapter 6).

Thus far nothing has been said about the physical properties of fly ash as they apply to the properties of concrete. Two important considerations that must be stressed here deal with its fineness and density. The maximum amount retained on the No. 325 (45 μm) sieve for both Classes has been set at 34% [3]. This amount, in the author's opinion, represents that portion of the fly ash that will exhibit very little pozzolanic reaction during the first 28 days of the life of the fly ash treated portland cement concrete. In addition (and still the author's opinion), the total of the percentage of iron oxide and L.O.I. represents that portion of ash that merely "goes along for the ride" and contributes nothing to the pozzolanic activity of the ash. So, if the user of fly ash in concrete is looking for some means of quality estimation of the added mineral, these three properties should be considered.

The density, or specific gravity, of fly ash, which is somewhat related to its morphology, is also a determining factor in its pozzolanic activity. A reproduction of a highly magnified photograph of a fly ash is shown in Figure 7-1. The structural configuration of an ash is basically made up of hollow spheres which have a "skin" of glassy, or amorphous silicon dioxide, or possibly complex compounds of silicon dioxide. This glassy skin (or what the author often refers to as the "egg-shell") is the principal reactive component in the fly ash. Small particles are generally more glassy than large particles because of their more rapid rate of cooling. Gases in the effluent from the coal combustion process, such as carbon dioxide, steam, and oxides of sulfur and nitrogen act in such a way as to bloat the semi-fluid particles into the spherical shape. The density of the spheres depends upon their size, i.e., the smaller the size, the higher the density. Regardless of their size, the glassy hollow spheres tend to reduce the density of the ash, as measured by standard methods [5] [6].

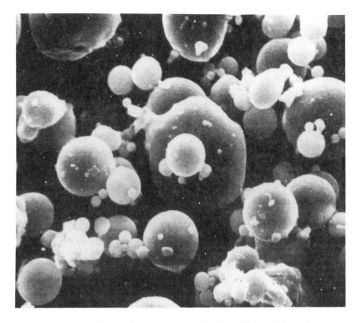

Figure 7-1. Microphotograph of a Class F fly ash.

Most fly ashes, whether they be of the Class F or C variety contain some crystalline material, the major portion of which is the mineral mullite, $3Al_2O_3 \cdot 2SiO_2$. It is the author's thinking that the crystalline component of fly ash, whatever its composition might be, can be considered non-reactive with respect to its pozzolanic response in portland cement concrete and that its addition is essentially equivalent to adding finely divided sand. For example, two Class F fly ashes having essentially the same amount of total silicon dioxide but different amounts of glassy silicon dioxide, are shown in Table 7-3. The amount of glassy silicon dioxide in each of the two ashes was determined by a previously published method [7]. Although both fly ashes met the requirements of the pozzolanic index test with lime, that of fly ash No. 1 exceeded that of No. 2 by 150% [3]. Fly ash No. 1 and 2 (Table 7-3) had densities of 2.31 and 2.56, respectively, and both contained essentially the same amount of iron oxide ($11-12\%$ Fe_2O_3). so, another general rule that the user of fly ash should consider says that the lower the density of the fly ash, the higher its pozzolanic activity is apt to be.

The amount of glassy, reactive material present in fly ash depends

Table 7-3. Silicon Dioxide Content of Two Class F Fly Ashes.

FLY ASH NO.	TOTAL SILICON DIOXIDE—%	GLASSY SILICON DIOXIDE—%
1	43.1	38
2	43.2	19

on at least two factors: (1) the temperature in the power producer's coal burning furnace and (2) the rate at which the ash is cooled. The higher the former and the quicker the latter, the greater will be the amount of glass in the ash. It should also be pointed out here that the fineness of fly ash is directly related to the fineness to which the coal is ground or pulverized prior to burning. While the pozzolanic activity of a given fly ash can be related to its density, several other factors enter into this relationship, such as its L.O.I., iron oxide and glassy silicon dioxide contents and fineness (or surface area). In recognition of the fact that fineness and density are inter-related, the suppliers of fly ash often express its overall fineness in units of cm^2/cm^3, which is the result of multiplying its fineness (cm^2/g) by its density (g/cm^3). Another relationship that tends to complicate the picture is that of fineness and L.O.I. In general, as the fineness of the fly ash decreases, its L.O.I. will also decrease. There is an optional physical requirement for Class F fly ashes, called the multiple factor, which is calculated as the product of L.O.I. and the amount retained on the No. 325 (45 μm) sieve. It was estimated in 1986 that over 50% of the ready mixed concrete in the U.S. contained fly ash [8]. Its use over the subsequent years has increased, but no reliable figures are available because of the concrete producers' reluctance to admit its use.

RAW/CALCINED NATURAL POZZOLANS—COMPOSITION AND PROPERTIES

Raw or calcined natural pozzolans, whose specifications are also included in ASTM C618 and methods of test are described in ASTM C311, are essentially of volcanic origin [3][5]. Some examples were cited earlier, but missing from that list were such things as certain diatomaceous earths, opaline cherts, shales, tuffs and clays. Most of these have to be activated by calcination (heating to a high temperature) followed by grinding to increase their fineness. After processing, most of the finished product is glassy in nature, but unlike the glassy component of fly ash, is not spherical in shape. This type of pozzolan has been assigned the designation of Class N (for natural). For comparison purposes, the chemical requirements for this class of pozzolanic material have been included in Table 7-1. The requirements for the Class N are essentially the same as those established for the Class F fly ash except for the loss on ignition. Very little of this class of pozzolan is used as a mineral admixture in modern concrete.

The author's experience with Class N pozzolans has been limited to a study of perlite. Perlite is a naturally occurring volcanic lava which contains entrapped water in its glassy and/or crystalline structure. Geologists chose the name, perlite, because under high magnification, the ore seems to be composed of tiny clusters of pearls. In its naturally occurring form, it has found little or no industrial use. However, after processing, it is widely used in the construction and other industries. These uses include (1) light weight concrete aggregate, (2) acoustical tile, (3) pipe insulation, (4) roof insulation board, (5) filter aids, (6) fertilizer carriers, (7) fillers, and (8) various unspecified industrial uses.

The processing of perlite involves a number of steps: (1) crushing the ore to sand sizes, (2) drying, (3) screening, (4) blending the various sand sizes, (5) heating to 1500° F to 2000° F in either a horizontal or vertical type furnace, and finally (6) removal of the treated fines in a cyclone separator. During the thermal treatment, the entrapped water is driven off and it expands (much like popcorn) 4 to 20 times its original volume and the finished product is snow white in color. Neither expanded perlite, expanded fines or expanded bag house dust seem to exhibit pozzolanic activity. However, after grinding the expanded perlite to a fineness of 10,000 cm^2/g, it does possess a high level of pozzolanic activity in concrete. Under the microscope, the ground expanded material appears to be made up of tiny, almost transparent, glass-like slivers. The composition of expanded perlite (typical, world-wide) is shown in Table 7-4.

GRANULATED BLAST FURNACE SLAG

Returning to artificial, or man-made, pozzolans, next consider granulated blast furnace slag which is a non-metallic product consisting essentially of silicates and aluminosilicates of calcium and other bases that is formed in a molten condition simultaneously with iron in a

Table 7-4. Typical Chemical Composition of Expanded Perlite.

CHEMICAL COMPONENT	%—BY WEIGHT	CHEMICAL COMPONENT	%—BY WEIGHT
SiO_2	75.1	Alkali (as Na_2O)	3.2
Al_2O_3	14.1	TiO_2	4.0
MgO	0.1	L.O.I.	1.2
CaO	0.4	SO_3	0.1
Fe_2O_3	0.9		

blast furnace and which is rapidly chilled, as by immersion in water [9]. The very rapid cooling of the slag results in most of its components being glassy (amorphous) which have a high pozzolanic activity. Blends of it with portland cement generally possess properties equivalent to or superior to plain portland cement after 3 to 7 days, at normal temperatures.

In the formation of the slag, the blast furnace is first charged with coke (carbon), iron ore and limestone ($CaCO_3$) and the mixture is heated to about 2600° F.

$$3C + 2Fe_2O_3 \rightarrow 3CO_2 + 4Fe \qquad (7\text{-}1)$$

The iron oxide component of the ore is reduced to molten iron. The limestone acts as a flux in the mixture and slowly floats to the top of the molten iron. The high temperature of the blast furnace converts it to calcium oxide which then combines with the silicon dioxide and aluminum oxide (in the ore and in the limestone itself) to form a complex mixture of calcium silicates and aluminates. The production of 1 ton of raw, or pig, iron requires, on the average, 1.7 tons of iron ore, 0.9 tons of coke and 0.4 tons of high grade limestone. Roughly, 0.5 tons of slag is produced per ton of raw iron. Although the chemical composition of blast furnace slag will vary, as it does in the case of other pozzolanic materials, a typical chemical analysis is illustrated in Table 7-5.

Although ASTM has defined a wide variety of cements created by the interblending of pozzolans such as fly ash and granulated blast furnace slag with portland cement, the author has taken the simple approach and states that a portland blast furnace slag cement is one that consists of a blend of the two, produced by intergrinding or blending, in which the slag constituent is between 25% and 70% of the weight of the finished product. When a type I portland cement is

Table 7-5.   Typical Chemical Composition of Blast Furnace Slag.

CHEMICAL COMPONENT	%—BY WEIGHT	CHEMICAL COMPONENT	%—BY WEIGHT
CaO	38.8	MgO	10.4
SiO_2	36.4	SO_3	1.2
Al_2O_3	9.6	Alkalies (as Na_2O)	0.6
Fe_2O_3	0.7		

involved, the finished product is referred to as a I-S cement. When a type I portland cement and a fly ash are combined, either by intergrinding or blending, the final product is called a type I-P cement.

Most of the granulated blast furnace slag finds its way into concrete via the type I-S cement. Only a relatively small amount of the available fly ash is sold by the cement producer, as the type I-P cement. The use of a type I-P cement vs. the direct addition of the fly ash to concrete has a number of advantages and disadvantages. The main advantage to the concrete producer who uses a type I-P cement is that he knows that the cement producer exerts stringent quality control of the fly ash prior to its being incorporated within the type I cement, thus assuring him of consistent performance in his concrete. The advantage realized by the concrete producer by using the pozzolanic material as a mineral admixture at the ready mix plant lies in versatility. In most cases, the concrete producer is not limited to the amount of pozzolan that he can add to his concrete, as long as the concrete meets job performance specifications, as he is when using a portland pozzolan cement.

One other advantage of using a portland pozzolan cement over the addition of the pozzolan directly to the concrete is the assurance that the portland cement and pozzolanic admixture are thoroughly blended. This is not always the case when the pozzolan is added as such a mineral admixture to concrete. The author has seen clumps, or agglomerates, of fly ash come down the chute of a concrete ready mix truck as large as tennis balls.

SILICA FUME

Another artificial, or man-made, pozzolan that appears to have tremendous potential in the concrete industry is silica fume. It is referred to as "fume" because it is so finely divided that its particle size approximates that of the solid particulates in smoke. Some purists prefer to call it microsilica and others as condensed silica fume. The silica fume designation will be used henceforth in this treatise. Silica fume is the by-product of the manufacture of elemental silicon and ferrosilicon alloys. During the process, which is carried out in an electric arc furnace (up to 3600° F), quartz (SiO_2) is reduced to elemental silicon and gaseous silicon monoxide. The latter is oxidized back to SiO_2 at the top of the open furnace to silicon dioxide when it comes in contact with the oxygen in the air.

$$SiO_2 + C \xrightarrow{\text{(HEAT)}} Si + CO_2 \text{ main reaction} \tag{7-2}$$

$$3SiO_2 + 2C \xrightarrow{\text{(HEAT)}} Si + 2SiO + 2CO_2 \text{ secondary reaction} \tag{7-3}$$

$$2SiO + O_2 \rightarrow 2SiO_2 \text{ secondary reaction (silica fume)} \tag{7-4}$$

Because of its extreme fineness, the silica fume has to be separated from the furnace effluent gases by a sophisticated dust collecting apparatus. The silicon dioxide derived from this process is a solid of extremely small particle size, about one-one hundredth that of a typical type I portland cement and consists of glassy spheres. The average diameter of the particles in silica fume has been established to be 0.1 μm (1 μm = 4×10^{-5} in.). Silica fume contains 86 to 98% silicon dioxide and because of its extreme fineness and high glassy SiO_2 content is a highly reactive pozzolanic material. Although the particle size of silica fume is near or above the upper limit for true colloids, its amorphous nature and specific surface area (in.^{2}/in.^{3} volume) suggests that its behavior is more closely related to a colloid than a glass. Published data indicate that there is little health hazard potential from the inhalation of amorphous silica fume due to its noncrystalline nature [10]. However, it is strongly recommended that dust masks and proper ventilation facilities be available when handling the dry material.

Current concrete practice is to limit the amount of silica fume used in concrete to 10 to 15 pounds per 100 pounds of cement, mainly because of economic reasons. While silica fume was once considered a waste material, its potential use in concrete has resulted in an escalation of its cost which now exceeds that of portland cement. The chemical composition of a typical silica fume is illustrated in Table 7-6.

The chemical requirements of silica fume proposed by Task Group

Table 7-6. Chemical Composition of a Typical Silica Fume.[a]

CHEMICAL COMPONENT	%—BY WEIGHT	CHEMICAL COMPONENT	%—BY WEIGHT
SiO_2	93.0	MgO	0.6
Al_2O_3	0.4	SO_3	0.3
Fe_2O_3	0.8	Alkali (as Na_2O)	0.96
CaO	0.6		

[a]The author has often referred to the composition of such materials as portland cement, fly ash, natural pozzolans, blast furnace slag, and now silica fume, as being typical. Just what is typical depends upon the reader, and the author admits that these typical compositions may not represent the materials with which one is accustomed.

4 of ASTM Subcommittee CO9.03.10 are summarized in Table 7-7. The chemical and physical properties of silica fume have been described in numerous publications and the author recommends one of these as supplementary reading [11]. Because of its extremely small particle size and hence its exceptionally high surface area, the method of Blaine is not practical for determining the fineness of silica fume. Through the use of nitrogen adsorption techniques, the fineness of silica fume has been estimated to be 20,000 m^2/kg (vs. a type I cement whose Blaine fineness is 300–350 m^2/kg).

PROPERTIES OF POZZOLAN—PORTLAND CEMENT CONCRETE

Pozzolanic mineral admixtures are used in concrete for a number of reasons. When the concrete is in the fresh state they improve its workability, although the finishers often complain about the stickiness of the concrete because the admixture increases the cohesiveness of the cementitious mass. Because they are generally finer than portland cement, their presence tends to eliminate, or minimize, bleeding (bleeding was discussed in Chapter 3) and reduce segregation. In the hard-

Table 7-7. Proposed (but Tentative) Chemical Requirements for Silica Fume.

CHEMICAL COMPONENT	%—BY WEIGHT
SiO_2—min %	85
Moisture content—max %	3.0
Loss on Ignition—max %	6.0
Available alkalies, as Na_2O—max %[a]	1.50

[a]Supplementary and optional requirement.

ened state, concrete containing any one of the types of pozzolan exhibits greater than normal strength (compressive, tensile and flexural), modulus of elasticity, resistance to sulfate attack, the deleterious alkali-aggregate reaction and (if air entrained) to freeze-thaw deterioration.

The author will first consider, from a rather simplistic viewpoint, the compressive strength increase of hardened concrete that is realized through the use of a pozzolanic mineral admixture. When portland cement and water interact, a number of products are formed (Chapter 1), but only two of these are important to this discussion, calcium hydroxide ($Ca(OH)_2$) and the calcium silicate hydrate. The latter is the "glue" that holds the paste, mortar and concrete together and often given the designation, CSH (Chapter 1). When a pozzolan is present in the mix, the $Ca(OH)_2$ produced by the hydrating portland cement phases (principally the calcium silicates) react with the active, or glassy, component of the pozzolan, most of which is silicon dioxide, to form more glue (CSH). These two aforementioned chemical reactions are simply expressed in Equations 7-5 and 7-6.

$$\text{Portland Cement} + \text{Water} \rightarrow \text{CSH (``Glue'')} + Ca(OH)_2 \quad (7\text{-}5)$$

$$Ca(OH)_2 + \text{Pozzolan} + \text{Water} \rightarrow \text{CSH (``Glue'')} \quad (7\text{-}6)$$

Equation 7-6 represents the pozzolanic reaction that is often referred to in the literature as well as in this text. The possibility that the alumino silicates present in some pozzolans participate in the pozzolanic reaction is not considered here.

POZZOLANIC REACTION—SEMI-QUANTITATIVE APPROACH

In order to quantify the nature of the chemical reactions expressed in Equations 7-5 and 7-6, a number of assumptions must be made.

1. A typical type I portland cement contains 52% tricalcium silicate and 20% dicalcium silicate and these two phases are the only source of calcium hydroxide in the concrete system.
2. The only products of the reaction between the two calcium silicate phases in the cement are calcium hydroxide and CSH (arbitrarily given the formula $3CaO \cdot 2SiO_2 \cdot 3H_2O$).

If these assumptions are acceptable to the reader, the amounts of water needed by 100 lbs of the cement and the quantities of the two products can be calculated stoichiometrically, as shown in Equations 7-7 and 7-8. The total $Ca(OH)_2$ and CSH produced then becomes 29.6 lbs and 58.9 lb per 100 lbs of cement, respectively, when the silicate phases in the portland cement are *completely hydrated* and 16.5 lb of water are needed for the reaction to go to completion.

$$
\begin{array}{ll}
52.0 \text{ lb} & X \text{ lb} \\
2[3CaO \cdot SiO_2] \; + \; 6H_2O \; \rightarrow & \\
\text{M.W. } 456 & 108 \\
Y \text{ lb} & Z \text{ lb} \\
3CaO \cdot 2SiO_2 \cdot 3H_2O \; + \; 3Ca(OH)_2 & \qquad (7\text{-}7) \\
342 & 222
\end{array}
$$

$$X = \frac{52 \times 108}{456} = 12.3 \text{ lb of } H_2O$$

$$Y = \frac{52 \times 342}{456} = 39.0 \text{ lb of CSH}$$

$$Z = \frac{52 \times 222}{456} = 25.3 \text{ lb of } Ca(OH)_2$$

$$
\begin{array}{ll}
20.0 \text{ lb} & X \text{ lb} \\
2[2CaO \cdot SiO_2] \; + \; 4H_2O \; \rightarrow & \\
\text{M.W. } 344 & 72 \\
Y \text{ lb} & Z \text{ lb} \\
3CaO \cdot 2SiO_2 \cdot 3H_2O \; + \; Ca(OH)_2 & \qquad (7\text{-}8) \\
342 & 74
\end{array}
$$

$$X = \frac{20 \times 72}{344} = 4.2 \text{ lb of } H_2O$$

$$Y = \frac{20 \times 342}{344} = 19.9 \text{ lb of CSH}$$

$$Z = \frac{20 \times 74}{344} = 4.3 \text{ lb of } Ca(OH)_2$$

In order to complete the picture two last assumptions must be made, i.e., 100 lb of a given pozzolanic mineral admixture contains 40 lb of active, glassy silicon dioxide and that the pozzolanic reaction product is also CSH.

$$\begin{array}{ccc} 40.0\ \text{lb} & X\ \text{lb} & Y\ \text{lb} \\ 2SiO_2 & + \ 3Ca(OH)_2 \ \rightarrow & 3CaO \cdot 2SiO_2 \cdot 3H_2O \quad (7\text{-}9) \\ \text{M.W.}\ 120 & 222 & 342 \end{array}$$

$$X = \frac{40 \times 222}{120} = 74.0\ \text{lb of Ca(OH)}_2$$

$$Y = \frac{40 \times 342}{120} = 114.0\ \text{lb of CSH}$$

Another exercise in stoichiometry follows in Equation 7-9. If 100 lb of the cement chosen as the example produces 29.6 lbs of $Ca(OH)_2$ as shown in Equations 7-7 and 7-8, then only 40 lb of the example pozzolan would be needed per 100 lbs. of cement in order to consume all of the liberated $Ca(OH)_2$. This means that a concrete having a C.F. of 400 lb (of the chosen cement) would theoretically require 160 lb of the chosen pozzolan in order to consume all of the $Ca(OH)_2$ generated by the former, when the chemical reactions are complete. With regard to the previously described chemical reactions, the following should be remembered:

1. Not all cements and pozzolans are alike in composition.
2. The stoichiometric calculations were based on the complete hydration of the cement.
3. The pozzolanic reaction only involves the reaction between the active silicon dioxide in the pozzolan.
4. Sufficient water is present in the concrete during its life to permit the reactions to go to completion.

It is important to note that 100 lbs of the cement chosen for the example in Equations 7-7 and 7-8 produces a considerable amount of calcium hydroxide. In the absence of a pozzolanic mineral admixture which interacts with it, the calcium hydroxide can exert several negative effects on the properties of the hardened concrete. First, with time and alternate wetting and drying it is leached from the interior of the concrete and becomes the principal source of efflorescence, a white, unsightly stain on the surface of the concrete. Second, that portion not leached out will react with carbon dioxide in the permeating atmosphere and be converted to calcium carbonate. This chemical conversion has a negative effect on the volume of concrete and leads to what is often called "carbonation" shrinkage. Third, its presence in concrete can aggravate sulfate attack (to be discussed later in this chapter). And, fourth, it has a pronounced influence on the bond strength between the paste and the aggregate in the concrete because it tends to deposit in the water pockets between the two. This phenomenon is pictured in Figure 1-5 wherein the plate-like crystals of $Ca(OH)_2$ are clearly shown adjacent to an aggregate socket in concrete at the age of 28 days. In the presence of a pozzolan, those crystals are absent (Figure 7-2) and replaced by fibrous CSH. Some might claim that $Ca(OH)_2$ contributes to the strength of concrete, but the author strongly disagrees with this concept.

Compressive Strength

The strength of concrete can be compared to that of a three link chain, the weakest link of which determines its strength. All of the links are important, as pictured in Figure 7-3, but the bond between the paste and the aggregate (the transition zone) is probably the most important.

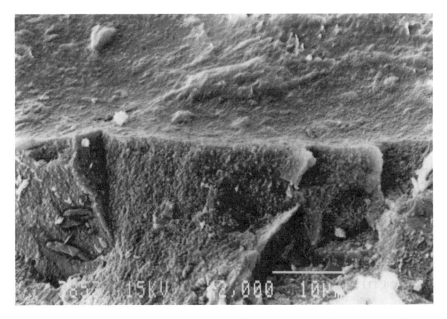

Figure 7-2. Microphotograph showing plate-like crystals of calcium hydroxide replaced by fibrous CSH in pozzolanic concrete bridging the gap between concrete matrix and aggregate.

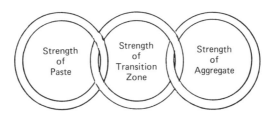

Figure 7-3. Three link chain concept of concrete strength.

Calcium hydroxide weakens this bond, or zone, and hence this particular link.

During the last eight years of the authors research career, he devoted a major portion of his time evaluating various pozzolanic mineral admixtures in *concrete*. It is difficult, if not impossible, to relate the results of the pozzolanic activity index test, with lime or portland cement, as described in ASTM C311, to those that might expected in concrete [5]. It became quite evident early in the author's concrete work that a ratio of fine to coarse aggregate had to be established which would yield a product of acceptable plastic properties at high and low cement factors with small and high amounts of pozzolan present. It was important that such a ratio be established and be used throughout the evaluation program so as to eliminate one important variable, so often introduced by other investigators which tends to cloud the results of their work.

It was found, for example, using the available laboratory coarse aggregate (3/4″, top size) and fine aggregate (fineness modulus = 2.65) that a 40:60 mix (40 pts. sand to 60 pts. stone, by weight) produced a very harsh and unworkable concrete at low cement factors, regardless of the amount of pozzolanic admixture in the mix. When the weight ratio of the two was changed to 50:50, concrete of low C.F.s and those of relatively high pozzolan content were very sandy in appearance and too sticky for proper placement. After an evaluation of numerous trial mixes, a sand to stone weight ratio of 45:55 was chosen for the concrete work.

In those concretes where a pozzolan was to be added for evaluation, it was added as a replacement for the cement on a 20% level, but on a volume for weight basis. This choice was made to avoid the influence of paste volume as a variable in the analysis of the compressive strength data. Because of the unpredictable water demands of the pozzolan containing concretes, the actual measured yields often varied from the theoretical (27 ft^3/yd^3), but seldom by more than 4%. The amounts of the concrete components were usually adjusted for yield before any of the concrete data was analyzed.

Each of the concretes fabricated in this program was designed to have a 3 ± 1/2″ slump and sufficient concrete was batched to permit casting of 4″ × 8″ cylinders for compressive strength, at various ages, and time of setting measurements. In some cases, enough concrete was fabricated for freeze-thaw durability and change in volume measurements. The batching and subsequent handling and testing of the test concretes was done in accordance with standard practices [12]. It was pointed out in Chapter 2 how the Omega Index Factor concept could be applied to the measurement of the compressive strength contribution of portland cement to concrete, in units of psi/lb/yd^3 of concrete. This same approach was applied to the measurement of the relative compressive strength contribution of pozzolanic minerals to concrete. In that work, various pozzolans whose chemical and physical properties are summarized in Table 7-8 were used. The test concretes are described in Table 7-9.

The compressive strength of the eight (8) test concretes, at three ages of test, are plotted as a function of their Omega Index Factors in Figures 7-4, 7-5, and 7-6. The method used to calculate the compressive strength contribution of the portland cement and that of the various pozzolans is pictured in Figure 7-7 and is self-explanatory. The calculated values are summarized in Table 7-10. Because portland cements differ in their chemical and physical properties, as do pozzolanic materials, the contribution of the two to compressive strength can vary over a rather broad range, as shown in Table 7-11. His limited experience with natural pozzolans, granulated blast furnace slag and silica fume does not permit the author to cite a realistic strength contribution range for those particular mineral admixtures. It is important that the reader remember that the values cited in Tables 7-9 and 7-10 are for pozzolans used in concrete in which 20% by *weight* of the cement has been replaced by an equal *volume* of the pozzolan. When a given pozzolan is substituted at some other weight (cement) to volume (pozzolan) ratio, the compressive strength contribution of the latter is apt to vary.

Table 7-8. Chemical Composition of Pozzolans Used in the Comparison of Compressive Strength Contribution to Portland Cement Concrete.

CHEMICAL AND PHYSICAL PROPERTIES	COMPOSITION—% BY WEIGHT				
	CLASS N[a]	CLASS F	CLASS C	GBFS[b]	SF[c]
SiO_2	73.2	43.4	32.5	31.5	97.2
Al_2O_3	13.1	18.5	21.9	17.0	0.0
Fe_2O_3	0.8	29.9	5.1	0.5	0.0
CaO	1.3	4.3	27.4	37.0	0.0
MgO	0.3	0.9	4.6	10.8	0.0
Total Alkali (as Na_2O)	7.26	0.60	1.10	N.D.[e]	0.23
L.O.I.[d]	2.4	1.2	1.2	N.D.	1.8
SO_3	0.1	1.2	2.8	2.8	0.02
Fineness- cm^2/g	10,000+	4120	4380	4210	10,000+
Specific Gravity	2.43	2.63	2.38	2.89	2.21

[a]Ground expanded perlite.
[b]GBFS—Granulated blast furnace slag.
[c]S.F.—Silica fume.
[d]L.O.I.—loss on ignition
[e]N.D.—not determined

Resistance to Sulfate Attack

Although the effect of pozzolans on the compressive strength of concrete is often thought of as being the principal reason for their use, their influence on the resistance of the treated concrete to sulfate attack may be just as important. Sulfate attack has been defined as a "chemical or physical reaction, or both, between sulfates usually in soil or ground water and concrete or mortar, primarily with calcium aluminate hydrates in the cement-paste matrix, often causing deterioration." [13]. At least three factors favor the resistance of concrete to sulfate attack: (1) the use of a portland cement having a low aluminum oxide content and hence a low tricalcium aluminate content (this is why a type V portland cement is referred to as a sulfate resistant cement), (2) a low calcium hydroxide content in the concrete, and (3) a low concrete permeability that can result from using a low water-cement ratio and/or a pozzolanic mineral admixture.

When portland cement comes in contact with water, the tricalcium aluminate phase immediately reacts with it and the interground cal-

Table 7-9. Concretes Used to Determine the Relative Strength Contributions of Various Pozzolans to Concrete.

CONCRETE COMPONENTS[a] AND PROPERTIES	1	2	3	4	5	6	7	8
Cement	552	479	403	449	442	445	446	445
Fine Aggregate	1410	1450	1475	1405	1400	1405	1385	1400
Coarse Aggregate	1740	1790	1815	1715	1710	1715	1695	1710
Pozzolan	—	—	—	78	92	82	74	98
				Fly Ash Class C	Fly Ash Class F	Natural Pozzolan Class N	Silica Fume	Ground Granulated Blast Furnace Slag
Water	294	296	295	300	310	318	303	300
Slump-in.	3-1/4	3-1/2	3-1/2	2-3/4	3	3-1/2	3	3
Air—%	1.6	1.4	1.5	1.6	1.2	1.4	1.8	1.5
W/C	0.53	0.62	0.73	0.67	0.70	0.71	0.68	0.67
O.I.F.	1041	773	552	670	631	627	656	664
Compressive Strength— PSI								
1-day	1325	1145	1005	1275	1170	1300	1350	1310
7-days	3750	3110	2620	3610	3555	3840	3990	3468
28-days	5790	4790	3870	5825	5315	6705	7120	6430

[a]Components—lb/yd^3.

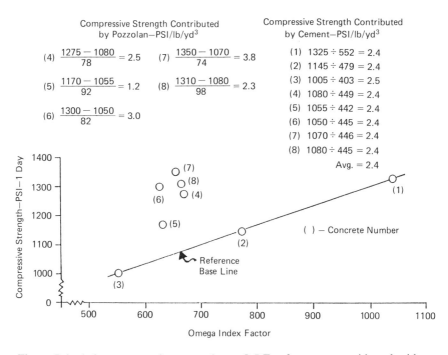

Figure 7-4. 1-day compressive strength vs. O.I.F. of concretes, with and without five different pozzolans, and a calculation of their relative strength contributions.

cium sulfate to form the compound called ettringite (see Equation 1-1). This is a positive volume change reaction, but the concrete is sufficiently plastic to accommodate the expansion without disruption. When the available (or soluble) sulfate ions in the concrete aqueous phase starts to decrease, the ettringite begins to react with more tricalcium aluminate to form the monocalcium aluminosulfate, or the low form of ettringite (see Equation 1-2). This results in a negative volume change, but again the concrete mass is still plastic enough to tolerate the dimensional change. In all probability, both of these reactions proceed almost simultaneously. Any tricalcium aluminate that remains after these two reactions go to completion reacts directly with water to form a tricalcium aluminate hexahydrate, $C_3A \cdot 6H_2O$. It is the low form of ettringite and the $C_3A \cdot 6H_2O$ that are particularly susceptible to sulfate attack. The iron containing analogues of the tricalcium aluminate reaction products are much more resistant to the effect of the ingressing sulfate ions and that is why a relatively high C_4AF content can be tolerated in a type V portland cement.

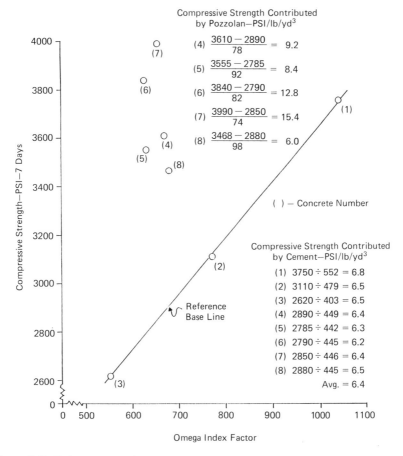

Figure 7-5. 7-day compressive strength vs. O.I.F. of concretes, with and without five different pozzolans, and a calculation of their relative strength contributions.

Next consider the sequence of events that occurs when concrete is exposed to ground water, sea water or soil that contain soluble sulfate ions, usually as alkali metal or magnesium sulfate ions, both of which (the cation and anion) diffuse into the concrete. First, when the calcium hydroxide, produced by the hydrating silicate phases of the cement, comes in contact with the intruding sulfate ions it is converted to calcium sulfate. This, in itself, can be a slightly expansive depending upon the amount of water present and the nature of the hydrated sulfate ion in the calcium sulfate that is formed. Second, the newly formed calcium sulfate then reacts with the tricalcium alumino

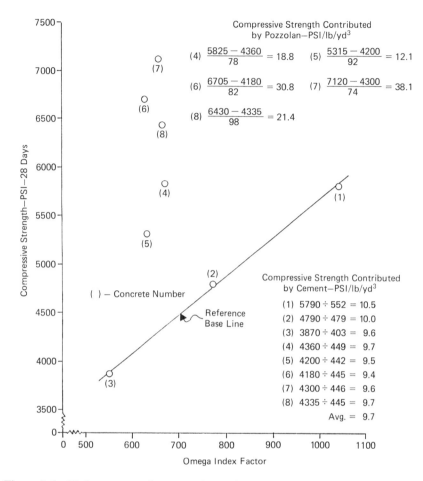

Figure 7-6. 28-day compressive strength vs. O.I.F. of concretes, with and without five different pozzolans, and a calculation of their relative strength contributions.

monosulfate and the $C_3A \cdot 6H_2O$ to form ettringite. Both of these reactions result in expansion because the products have a volume approximately 150% that of the reactants. The products of all these reactions form crystals in the pores and voids in the concrete which ultimately destroy the integrity of the concrete by their expansive forces.

The presence of carbon dioxide in the permeating atmosphere can also complicate the sulfate attack picture as a result of the formation of the mineral thaumasite which is a complex calcium carbonate (formed by its reaction with calcium hydroxide)-silicate-sulfate hydrate which

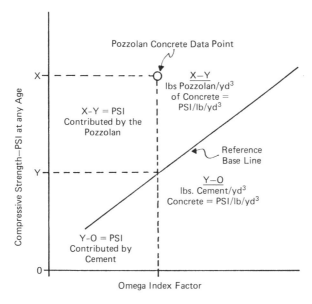

Figure 7-7. Method used to calculate the compressive strength contribution of cement and pozzolan to concrete in PSI/lb/yd³ of concrete.

has a larger volume than the compounds from which it is formed. When all of the previously described expansive reactions that account for sulfate attack and the subsequent disruption and destruction of concrete are considered, two key issues make themselves very clear: the amount of calcium hydroxide in the concrete and its permeability. This is where the addition of pozzolanic mineral admixtures to concrete comes into play.

By virtue of the pozzolanic reaction, most of the calcium hydroxide in concrete is converted to stable calcium silicate hydrate, CSH, which has a number of significant consequences. First, its near absence in the pozzolan containing concrete minimizes any formation of calcium sulfate by the intruding sulfate ions from its environment and thus reducing the possibility of ettringite and/or thaumasite formation. Second, the solubility of calcium hydroxide permits water to leach it out from the concrete (efflorescence), leaving empty voids and capillaries, which increases its permeability. The pozzolanic reaction prevents this from happening and as a result the concrete is less permeable, less susceptible to sulfate ion attack and has an increased strength. In this connection, it has been reported that a less compact

Table 7-10. Summary of Compressive Strength Contributions Made by the Portland Cement and Pozzolans Illustrated in Figures 7-4, 7-5, and 7-6.

AGE OF TEST—DAYS	COMPRESSIVE STRENGTH CONTRIBUTION—PSI/lb/yd³ CONCRETE					
	CLASS F FLY ASH	CLASS C FLY ASH	CLASS N POZZOLAN[a]	GBFS[b]	SILICA FUME	CEMENT TYPE I
1	1.2	2.5	3.0	2.3	3.8	2.4
7	8.4	9.2	12.8	6.0	15.4	6.4
28	12.1	18.8	30.8	21.4	38.1	9.7

[a]Ground expanded perlite
[b]GBFS—Granulated blast furnace slag

CSH was found in well hydrated portland-pozzolan cements [14]. This would have the effect of filling up the voids and capillaries. Third, under stress, microcracks tend to form at the paste-aggregate transition zone (Figure 7-3) in concrete. This zone plays an important role in determining the mechanical properties and long time durability characteristics of concrete. In plain portland cement concrete, the transition zone is generally much less dense than the paste and aggregate adjacent to it and contains a large amount of plate-like crystals of calcium hydroxide (Figure 1-5). Therefore, it is prone to respond to tensile stresses and those created by normal thermal and humidity changes. The presence of a pozzolan in concrete reduces the thickness of the transition zone and the degree of orientation of any of the calcium hydroxide crystals that might be present.

The resistance of concrete to sulfate attack according to the nature of its cementitious constituent(s) can be rated as follows:

1. Type V cement + pozzolan (greatest)
2. Type V cement
3. Type II cement + pozzolan
4. Type II cement

Table 7-11. Contribution to Compressive Strength of Concrete.

AGE OF TEST—DAYS	CEMENT, TYPES I–V—PSI	FLY ASH, CLASS F AND C—PSI
1	1–3	0–8
7	5–8	7–15
28	8–12	9–20

5. Type I cement + pozzolan
6. Type I cement (least)

Determining where type IV and III cements, with and without pozzolans, fit into this scheme is beyond the bounds of the author's experience.

As to the rating of pozzolanic mineral admixtures with respect to their effect on sulfate resistance, a number of physical and chemical properties must be considered.

Physical—fineness and density
Chemical—% active SiO_2, % Al_2O_3, % CaO, and % Fe_2O_3

If only these parameters are considered, the various types of pozzolans listed in Tables 7-2, 7-4, 7-5, and 7-6 can be rated in the following order of decreasing effectiveness in deterring sulfate attack:

$$\frac{\text{Silica}}{\text{fume}} > \frac{\text{Class N}}{\text{pozzolan}} > \frac{\text{Granulated blast}}{\text{furnace slag}} > \frac{\text{Class F}}{\text{fly ash}} > \frac{\text{Class C}}{\text{fly ash}}$$

There is evidence that certain Class C fly ashes having a ratio, $\%CaO-5/\%Fe_2O_3$, called the sulfate resistance factor, R, of less than 0.50 will greatly improve sulfate resistance [15].

The results of sulfate resistance tests on concrete exposed to real-life conditions, over an extended period of time, are difficult to document. The results of long-term tests of silica fume concretes conducted in Oslo, Norway, are very impressive [16]. After 26 years of exposure, concretes containing a 15% replacement of a type I cement by silica fume have shown the same performance as similar concretes made from a type V cement. The water at the test site contained up to 6 grams of sulfate ion per liter and had a pH of 3. The magnitude of the sulfate ion content would rank the environment as being severe [17].

Sea water attack on concrete is really a form of sulfate attack, but other intruding ions enter the picture. Sea water contains approximately 3.3% soluble salts, by weight, which includes sodium chloride (2.7%), magnesium chloride (0.3%), magnesium sulfate (0.2%), and calcium sulfate (0.1%). There is also some dissolved carbon dioxide in the water. Again, the calcium hydroxide in the concrete as well as the concrete's permeability are important factors in determining its durability because the intruding ionic species react with the former in one way or another leading to expansive reactions.

Influence on Alkali-Aggregate Reaction

Another performance benefit realized through the use of pozzolanic mineral admixtures is their influence on the deleterious alkali-aggregate reaction that often takes place in plain portland cement concrete. Three types of alkali-aggregate reactions are currently recognized.

1. *Alkali-silica reaction*—which involves the interaction between the alkali in the concrete with certain aggregates containing active silicon dioxide (silica). Unreactive silica, such as quartz, has an orderly arrangement of its silicon-oxygen tetrahedra within its crystal structure, whereas the reactive forms are comprised of randomly arranged tetrahedral networks. Such aggregates as opal, chert and chalcedony react with alkali in the concrete (either introduced by the portland cement, admixtures or the environment) to form a gel on the surface of the aggregate which imbibes moisture causing the gel to swell. The swelling develops pressure on its surroundings and results in disruptive cracking.
2. *Alkali-carbonate reaction*—is caused by the presence of certain types of carbonate containing aggregate such as argillaceous dolomitic limestones that contain metastable calcium carbonate. These carbonates undergo a chemical reaction with alkali hydroxides creating products whose volume is larger than that of the reactants (Equation 7-10).
3. *Alkali-silicate reaction*—is the result of aggregates containing greywackes (sand stone containing feldspars or clays) and those that exfoliate, such as vermiculites, interacting with the alkali and producing expansive stresses within the mass of the concrete.

$$CaMg(CO_3)_2 + 2NaOH \rightarrow Mg(OH)_2 + CaCO_3 + Na_2CO_3 \quad (7\text{-}10)$$

The rate and extent of the alkali-aggregate reaction, regardless of the category in which it falls, depend on several factors, i.e., (1) total quantity of reactants, (2) temperature, (3) particle size of the alkali-active ingredient(s), (4) availability of the alkali and hydroxyl ions to the reaction sites which involves the characteristics of the pores and capillaries in the concrete paste and the availability of pore liquid as an ion transport medium. One of the first indications that any one or more of the previously described types of alkali-aggregate reactions are taking place is map cracking on the exposed surface of the concrete. Map cracking appears as intersecting cracks which are usually random, but on a fairly large scale that extend below the surface of the concrete and may reach a width of as much as 0.5 inches and are due to differential rates of volume change in different portions of the concrete. With the increase in (1) demand for portland cement concrete, (2) the depletion of non-alkali-reactive aggregate, and (3) the alkali content of cement, the rate of occurrence of the alkali-aggregate reaction has no way to go but up, unless the destructive reactions can be minimized, or eliminated, by the addition of an admixture that will serve as a means for the exorcism of the alkali in concrete.

The use of pozzolanic mineral admixtures seems to be the best solution to the aforementioned dilemma at this point in time. The influence of pozzolans on the normal alkali-aggregate is basically attributed to their (the pozzolan) fineness and high silicon dioxide content, most of which is glassy and very susceptible to alkali attack. Several theories have been proposed to explain how pozzolans negate the alkali-aggregate reaction. *One* theory holds that the alkalies in the concrete preferentially combine with the highly reactive silica in the pozzolanic particles, rather than the alkali active silicas, carbonates or silicates present in the aggregate. Because the pozzolanic particles are evenly distributed throughout the concrete, any swelling or expansion that occurs is so evenly distributed (as against local concentrations) that the expansive forces can be tolerated by the concrete. An artist's conception of this distribution of expansive forces is illustrated in Figures 7-8 and 7-9. Somewhat related to this thinking is the theory that a non-swelling sol, or liquid gel made up of lime, alkali and silica (from the pozzolan) is formed in place of the normal expanding alkali-silicate gels. A *second* school of thought holds that the nature of the CSH produced by the pozzolanic reaction has a lower CaO-SiO_2 ratio than that produced by the hydrating cement silicates and thus is able to incorporate large amounts of alkali ions in its structure, leading to a reduction in the amount available for the alkali-aggregate reaction. A *third* theory simply says that the presence of the finely divided pozzolans in the concrete restricts the mobility of the alkali and hy-

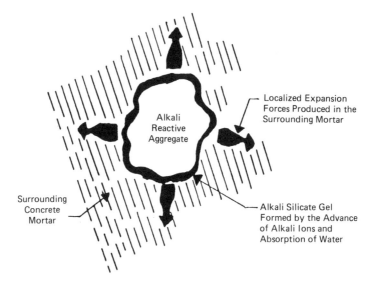

Figure 7-8. Alkali reactive aggregate in plain concrete.

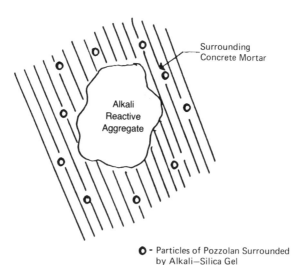

O - Particles of Pozzolan Surrounded by Alkali–Silica Gel

Figure 7-9. Preferential alkali attack on finely divided pozzolan—due to its large specific surface.

droxyl ions needed to cause the destructive expansive reactions. The author has no special feelings for any one of the proposed mechanisms and is satisfied with the idea that any combination of these could be at work.

Freeze-Thaw Durability

Three other aspects of pozzolanic concrete durability should be mentioned here. *First,* the resistance to freeze-thaw deterioration of concrete containing pozzolanic mineral admixtures warrants some consideraton. It was pointed out in Chapter 6 that when plain air entrained concrete had sufficient compressive strength at the time of exposure to the hostile environment and if the air entrained in the concrete had a low spacing factor, the possibility of the concrete withstanding the rapid freezing and thawing exposure was very good (see Equation 6-5). The same factors appear to apply to air entrained pozzolanic concrete. Five examples are described in Table 7-12. As was the case with plain air entrained concrete, the cut-off point (pass or fail) of the F-T D.I. seems to be 450.

Shrinkage

Second, one of the often cited advantages of using pozzolans in concrete is that they reduce the shrinkage of concrete [18]. Although the W/C in pozzolanic concretes is generally higher than that of plain concretes, in order to attain the desired slump, a good part of that

Table 7-12. Freeze-Thaw Durability of Pozzolanic Concrete.[a]

CONCRETE NO.	C.F.— LB/YD[3b]	POZZOLAN CONTENT LB/YD[3c]	HARDENED AIR—%	SPACING FACTOR IN.	COMPRESSIVE STRENGTH PSI	F-T D.I.	PASS (P) OR FAIL (F)[d]
1	400	110	5.0	0.0079	3730	472	P
2	380	130	5.6	0.0068	3025	445	F
3	380	150	5.4	0.0071	2910	410	F
4	500	125	6.0	0.0068	3365	495	P
5	525	125	5.0	0.0080	4480	560	P

[a]ASTM C666, rapid freeze-thaw in water
[b]Type I cement
[c]Class F fly ash
[d]RDME, No. 1-78, No. 2-52, No. 3-52, No. 4-82, No. 5-93.

water is prevented by the pozzolanic reaction products from escaping during the course of drying. This results in the creation of fewer empty voids (pores and capillaries) in the concrete which is responsible for a good portion of concrete shrinkage. Therefore, there is good reason for believing that pozzolans will reduce the shrinkage of concrete and the published data indicates that the reduction may be as high as 25% [19]. Although it seemed rather academic to go through the exercise, the author elected to evaluate approximately fifty (50) different pozzolans with respect to their influence on the shrinkage of concrete. The reference concretes contained 500 lbs of various types and brands of cement/yd^3 of concrete and the pozzolanic concrete mixes were identical to the reference, except that 20% of the cement was replaced by an equal volume of the various pozzolans. The test method established in ASTM C157 was used in the evaluation [20]. The reduction in shrinkage resulting from the use of the pozzolan was found to be in the range of 14% to 18%. This is significant when the vagaries of the method of measurement are considered.

Finally, because concretes containing pozzolans are denser and less permeable, damage due to deicing slats is reduced. The lowered permeability tends to counteract the intrusion of chloride ions associated with the erosion of embedded iron and its alloys.

OBJECTIONS TO USE OF POZZOLAN ADMIXTURES

Effect on Time of Setting

One of the principal objections to the use of pozzolans in concrete is their influence on the time of setting of the concrete. The use of a pozzolanic mineral admixture usually extends the time of setting of the concrete giving rise to complaints from the finishers and contractors. There are logical reasons for the difference in time of setting of the two types of concrete. The time of setting of plain concrete has been shown in Chapter 2 to be a function of the W/C and the C.F. of the concrete and the time of setting of plain concrete was shown to be linearly related to the O.I.F. of the concrete (Figure 2-5). This same concept can be used to better understand the influence of pozzolanic admixtures on the time of setting of concrete [21].

Four plain concretes and four pozzolan containing concretes, all fabricated from the same portland cement, are described in Table 7-13; and the initial and final times of setting of the eight concretes are

Table 7-13. Concretes for Time of Setting Measurements With and Without Fly Ash.

CONCRETE NO.[a]	1	2	3	4	5	6	7	8
Cement[b]	527	529	522	577	416	421	399	420
Fly Ash	—	—	—	—	85[c]	89[d]	95[e]	75[f]
Water	273	291	305	318	320	281	347	365
Slump-in	2-1/2	3-1/4	3-3/4	5-3/4	4-1/2	3-1/2	3-1/4	3
W/C	0.52	0.55	0.58	0.62	0.77	0.67	0.87	0.87
O.I.F.	1013	962	900	834	540	628	459	483
Time of Setting—min								
Initial	261	280	280	296	363	347	310	260
Final	353	366	375	400	467	450	405	355

[a] All concrete components—lb/yd³, yield = 27.0 ft³/yd³.
[b] Type II portland cement
[c] Fly ash, Class F
[d] Fly ash, Class F
[e] Fly ash, Class C
[f] Silica fume

plotted as a function of their Omega Index Factor in Figure 7-10. The data points for six of the concretes fall on a straight line, but those representing concretes No. 7 and No. 8 fall significantly below the linear plots. The extended times of setting of concretes No. 5 and 6 are the results of what was done to the plain concrete to accommodate the addition of the two pozzolans, i.e., the C.F. was reduced and the W/C was increased, both of which favor a delay in time of setting. Therefore, their delay in time of setting is not the result of some chemical process, but only by the magnitude of the changes made in their C.F.s and W/Cs.

Concrete No. 7 exhibits initial and final times of setting comparable to those of a concrete having an O.I.F. of 800 with an accelerated time of setting amounting to 75 and 80 minutes, final and initial, respectively, over what might be expected if only the C.F. and W/C parameters were at work. This behavior leads to the conclusion that something chemical is occurring in the mix. The pozzolan in No. 7 is a Class C fly ash, some of which possess hydraulic properties producing relatively large amounts of ettringite which contributes to the setting of portland cement. The time of setting of Concrete No. 8, containing silica fume, is essentialy the same as that of the plain concrete, No. 1, from which it was derived. The ability of the silica fume concrete to off-set the set retardation effects of its higher W/C and lower C.F. have been observed by others [22]. One explanation

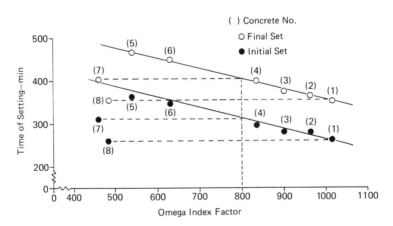

Figure 7-10. The influence of selected mineral admixtures on the time of setting of concrete as determined by using the Omega Index Factor.

for this is the rapid reaction of the calcium ions, liberated by the hydrating portland cement with the finely divided pozzolan [23]. This will result in an acceleration of the cement hydration as discussed in Chapter 2. On rare occasions the author has heard about (but never worked with) fly ashes that aggravate the set retardation over and above that normally expected, using the O.I.F. concept. This might be explained on the grounds that fly ash is often transported in tank trucks that previously contained other materials, some of which can act as strong retarders on the hydration of portland cement. The excessive retardation could thus be due to contamination.

Most of the author's work with pozzolans and their effect on time of setting has been centered on fly ash and silica fume, but the results of his rather limited experience with concretes containing natural pozzolans and ground granulated blast furnace slag seem to fall in line with those just described for concretes Nos. 5 and 6.

Bleeding/Curing

Another often voiced complaint by the concrete finishers about pozzolanic concretes has to do with the phenomenon known as bleeding and/or the lack of it, which was referred to earlier in Chapter 6. A concrete that exhibits little or no bleeding can undergo what is known as plastic shrinkage. This is caused by the rapid loss of moisture from the surface of fresh concrete due to environmental conditions such as low relative humidity, high temperature and excess wind velocity over the surface of the concrete. If the rapid loss of moisture on the surface is not counteracted either by bleed water or moist curing (fog spraying), tensile stresses develop on the surface and cracking results. These cracks can only be eliminated by reworking, or refinishing, the concrete surface. Since mineral admixtures tend to minimize bleeding, especially silica fume, because of its extreme fineness, the concrete must be adequately cured, which raises its cost as well as the hackles of both the finishers and contractors.

Discoloration

While discussing the use of pozzolanic mineral admixtures on the negative side, the problem of color that some can impart to the treated concrete must be broached. Fly ashes probably vary in color more than any of the pozzolans. The author has seen ashes as dark as the ace of spades and as light as vanilla ice cream in color and all of the shades of grey, brown and tan in between these two extremes. The color of fly ash will vary from source to source and often within a given source. The principal contributors to the color of a pozzolan are its iron oxide and carbon (loss-on-ignition) contents. To illustrate this last point, five fly ashes of known chemical composition were chosen for color comparison in Figures 7-11 and 7-12.

The three fly ashes shown in Figure 7-11 have a fairly constant iron oxide (reported as Fe_2O_3) content, varying only between 8.5% and 9.0%. However, there is a considerable difference in their loss-on-ignition values; i.e., 0.5% for sample A, 3.6% for B, and 6.3% for C. When the iron oxide content is reasonably constant, the effect of an increasing loss-on-ignition is a distinct darkening in color. The two ashes pictured in Figure 7-12 have the same loss-on-ignition, 0.7%, but sample D has 5.6% and E has 15.1% iron oxide, expressed as Fe_2O_3. The difference in color between the two illustrates the darkening effect of iron oxide on the color of fly ash when its loss-on-ignition is maintained at a constant level. While discussing color, it should be pointed out here that the grey color of portland cement is mainly due to its iron oxide content because its loss-on-ignition is

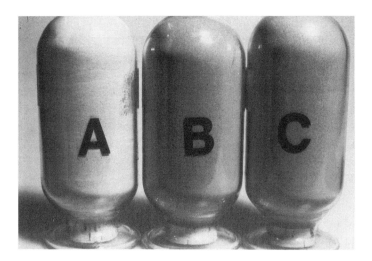

Figure 7-11. Variation in fly ash color due to carbon content.

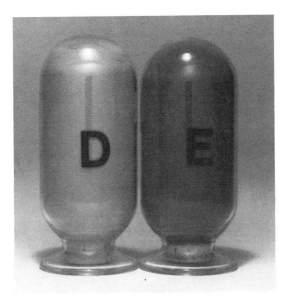

Figure 7-12. Variation in fly ash color due to iron oxide content.

usually less than 1% (most of which is due to moisture and carbon-ates). White cement has a very low iron oxide content.

Because of its low iron oxide and carbon contents, ground granu-lated blast furnace slag is somewhat lighter in color than portland cement and the blends of the two are usually only slightly lighter in color than portland cement. Concrete fabricated from blends of that pozzolan and portland cement often exhibits a bluish green color which fades away on exposure to light and air. The author has always at-tributed this to the presence of certain metal sulfides, sulfites or thio-sulfates formed in the pozzolanic concrete which are oxidized to less colorful compounds on exposure to the environment.

It was mentioned earlier that ground expanded perlite is as white as snow. When it is added to concrete in amounts exceeding 15%–20% by weight, on weight of cement, the resulting concrete is lighter in color than plain concrete.

The color of silica fume varies from light to bluish grey. Since its iron oxide and carbon contents are usually low ($Fe_2O_3 = 0.5\%-2.0\%$, loss-on-ignition $= 0.5\%-2.0\%$) that color must be due to either trace elements or its extreme fineness. When it is dispersed in water, to produce a slurry, the resulting mixture is very dark and often black in color. Again, this could be the result of an increase in fineness due to dispersion. For example, silver metal as we know it has a bright metallic luster, but in the finely divided state it is black.

Experience with concretes containing pozzolanic mineral admixtures has shown that when the pozzolan is present in amounts exceeding 20% by weight (on weight of cement) the color of the pozzolan is very apt to dominate the color of the concrete. This is not to say that a 20% addition will produce an inferior concrete, but can, in architectural concrete, cause aesthetic problems.

ECONOMICS OF POZZOLAN USE

Determining the economical advantage of using pozzolans in portland cement concrete is somewhat filled with vagaries and uncertainties because (1) the costs of portland cement and of the pozzolanic materials fluctuate periodically and differ from one geographic area to another, (2) the amounts of the two vary from one concrete producer to another, and (3) the pozzolan is used sometimes as a partial replacement for portland cement and at other times as a supplement to the cement. Probably the most valid measure of the economical advantage of using a pozzolan, as a partial replacement for portland cement in concrete can be illustrated by the concretes described in Table 7-14. Two different fly ashes were chosen as examples and have been used to replace 20% (by weight) of the cement in Concrete No. 1, on a volume for weight basis. Concrete Nos. 2 and 3 are additional reference concretes and will be used to establish the O.I.F. reference line. The compressive strength data, at two ages of test, are shown in Figures 7-13 and 7-14.

If the economic advantage (or disadvantage) of using either of the two fly ashes is going to be based on early compressive strength; i.e., one-day, the data in Figure 7-13 must be considered. The compressive strength of concrete No. 4, containing fly ash A, is the same as that of a plain concrete that has an O.I.F. of 965. Using an estimated W/C of 0.55 (to produce a slump of 3-1/2 ± 1/2"), that plain concrete would have to contain (965 × 0.55) 531 lb of cement/yd³. Assuming that the delivered cost (to the concrete producer) of the cement is $0.03/lb and of the fly ash is $0.015/lb, it follows then that the plain concrete would cost (531 × $0.03/lb) $15.93/yd³ and the cementitious materials in Concrete No. 4 would cost

Table 7-14. Concretes Used to Illustrate the Economics of Using Fly Ash in Concrete.

CONCRETE COMPONENTS[a]	CONCRETE NO.				
	1	2	3	4	5
Cement[b]	552	479	403	449	442
Fine Aggregate	1410	1450	1475	1410	1410
Coarse Aggregate	1740	1790	1815	1740	1740
Fly Ash	—	—	—	78[c]	92[d]
Water	294	296	295	300	310
Physical Properties					
Slump-in.	3-1/4	3-1/2	3-1/2	3	3
Air—%	1.6	1.4	1.5	1.6	1.2
W/C	0.53	0.62	0.73	0.67	0.70
O.I.F.	1041	772	552	670	631
Compressive Strength—psi					
1-Day	1325	1145	1005	1275	1170
7-Day	3750	3110	2620	3610	3555
28-Day	5290	4500	3870	5525	5315

[a]All components—lb/yd^3 of concrete.
[b]Type I cement.
[c]Fly ash A, Class F.
[d]Fly ash B, Class F.

$$449 \text{ lb cement} \times \$0.03/\text{lb} = \$13.47$$
$$78 \text{ lb pozzolan} \times \$0.015/\text{lb} = \underline{\$\ 1.17}$$
$$\$14.64$$

The difference in the costs of the two concretes, having the same early strength, represents a savings of ($15.93 − $14.64) $1.29/yd^3 of concrete. Although certain cost differences such as that of supplemental chemical admixtures, AEA's and aggregates have not been considered, they would be minor in magnitude.

Following the previous line of thinking, a plain concrete having the same one-day compressive strength as Concrete No. 5, containing fly ash B, would have to have an O.I.F. of 805. With an estimated W/C of 0.61 (to attain the same slump), the C.F. of the concrete would have to be 491 lb. This brings the cost of the concrete to (491 lb/yd^3 × $0.03/lb) $14.73/yd^3. The cost of the cementitious portion of Concrete No. 5 is

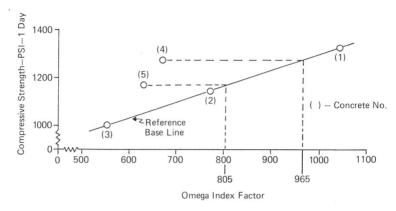

Figure 7-13. Estimation of the savings (or cost) in using pozzolans by the Omega Index Factor technique.

$$442 \text{ lb cement} \times \$0.03/\text{lb cement} = \$13.26$$
$$92 \text{ lb fly ash B} \times \$0.015/\text{lb of fly ash} = \underline{\$\ 1.38}$$
$$\$14.64$$

This represents a savings of only ($14.73 − $14.64) $0.09/yd^3 of

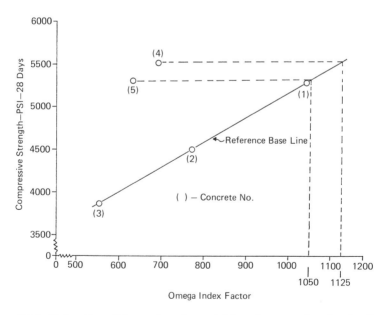

Figure 7-14. Estimation of the savings (or costs) in using pozzolans by the Omega Index Factor technique.

concrete and the *economic advantage* of using fly ash B, based on early strength is, at best marginal. The words *economic advantage* have been emphasized lest the performance advantages provided by the pozzolan such as protection against sulfate attack and the alkali-aggregate reaction be forgotten.

A different picture is painted when later compressive strengths; i.e., 28 days, are considered the criteria for the concrete. On examining the data in Figure 7-14 and applying the same type of reasoning used in analyzing the early strength data, one can calculate that the savings realized through the use of fly ash A and fly ash B would be $2.58 and $2.04/yd^3 of concrete, respectively. A summary of that analysis is given in Table 7-15.

The subject of economics pertaining to the use of silica fume warrants special consideration. Of all of the pozzolanic mineral admixtures thus far investigated, it is the most efficient; and when used in small quantities; i.e., 10–20% by weight, replacement for portland cement it has a potential strength contribution of 3 to 4 times that of portland cement (see Table 7-10). Although silica fume is commercially available in several forms, i.e., dry, densified dry and in a water-based slurry, the latter is preferred because of fewer handling and shipping problems. For example, the dry material has a bulk density of 12 to 15 lb/ft^3 (vs. 94 lb/ft^3 for portland cement). This means that a tank delivery truck can only carry 13%–16%, by weight, of that normally transported, in terms of portland cement. After densification, the volume of the silica fume is reduced by a factor of 2.5 to 3.5. The cost of transporting silica fume as a water-based slurry is much lower (on the basis of pounds of silica fume/gallon of slurry) but because the addition rate of the slurry (usually 50% solids) is greater, larger and more complex dispensing equipment must be employed than that used for the common chemical admixtures because

Table 7-15. Economics of Using Fly Ash A and Fly Ash B, Based on 28-Day Compressive Strength.

COST OF PLAIN CONCRETE—$/YD^3	COST OF FLY ASH A CONCRETE—$/YD^3	COST OF FLY ASH B CONCRETE—$/YD^3	SAVINGS $/YD^3
574 lbs cement × $0.03/lb = 17.22	14.64	—	2.58
556 lb cement × $0.03/lb = 16.68	—	14.64	2.04

the silica fume slurry must be agitated to prevent it from separating. Whether a pozzolan is added dry or as a slurry, a water reducing admixture, preferably of the high range type, should be used in order to realize its full benefit. This is especially true when fine pozzolans such as ground expanded perlite and silica fume are used. Although the cost of storage, agitation and dispensing equipment are usually provided by the supplier, the user can be assured that it is a part of the cost of the admixture.

While the aforementioned use of the O.I.F. to estimate the economics involved in the use of pozzolanic mineral admixtures is rather unorthodox, it seems to be a unique approach to unlocking the question of what the potential savings are when the pozzolan is used as a partial replacement for portland cement. The problem of estimating the economics of adding pozzolans as supplements to the cement in concrete is much more complex and can only be established by trial mixes and subsequent calculations based on total cost, properties and yield of the treated concrete vs. those of the plain concrete. The principal reason for using a pozzolanic mineral admixture as a supplemental cementitious material to concrete is to attain higher than normal strengths. In many cases those high levels of compressive strength cannot be reached by any other method.

HIGH STRENGTH CONCRETE

High strength concrete seems to have become the key word in today's concrete technology. In the early 1940s 4000 psi (at 28 days) was considered to be representative of high strength concrete. This level jumped to 7000 psi in the late 1950s and early 1960s. Concrete strengths of 14,000 to 19,000 psi is now being viewed as the criteria for high strength. Just how far we can go to reach an ultimate in strength in the future is anybody's guess. These abnormally high strength levels can only be realized by using high C.F.s, sturdy, well graded aggregates, active pozzolans and low W/Cs (through the addition of high range water reducing admixtures).

The compressive strengths, at four ages of test, of two concretes are shown in Figure 7-15. Both were fabricated from a type I portland cement and had a slump of 3-1/4″. The reference concrete contained a water reducing admixture but no silica fume. The second concrete contained both a high range water reducing admixture and silica fume.

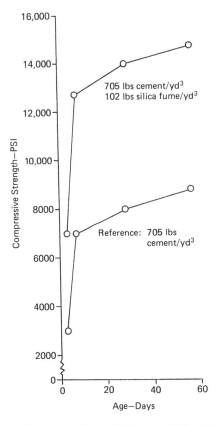

Figure 7-15. High strength concrete through the use of silica fume and a high range water reducing admixture.

The HRWR was of the naphthalene type and was added at the rate of 16 fluid oz/100 lb of total cementitious material. The numerical differences in the compressive strengths at the chosen ages of test are summarized in Table 7-16.

POZZOLAN FILLER EFFECT

Finely divided mineral admixtures can also exert what is often called the "filler effect," by simply decreasing the average size of the pores in the cement paste. This effect can also contribute to compressive strength in addition to that of its pozzolanic reaction products. Somewhat related to this, there has recently been a move in the U.S. to

Table 7-16. Compressive Strength Increase Exhibited by Silica Fume— HRWR Treated Concrete.

AGE-DAYS	INCREASE-PSI
3	4000
7	5750
28	6000
56	5950
	Avg: 5425

allow the cement producer to intergrind up to 5% limestone (on weight of portland cement) because of its strength enhancement when the cement is used in mortars and concrete. It was first thought that the added finely divided limestone contributed strength strictly through the filler effect. Now there is evidence that it can, in the absence of sufficient calcium sulfate to prevent the formation of C_4AH_{13}, react with the latter to form tricalcium carboaluminates analogous to ettringite and its monosulfate form which would account for the increase in sulfate resistance of the cement [24].

REFERENCES

[1] "Cement and Concrete Terminology," *SP-19, American Concrete Institute, Detroit, Michigan,* pg. 107 (1988).

[2] Lea, F. M., "The Chemistry of Cement and Concrete," Chemical Publishing Co., First American Edition, New York, NY, pp. 3–4 (1971).

[3] ASTM C618, "Standard Specification for Fly Ash and Raw or Calcined Natural Pozzolan for Use as a Mineral Admixture in Portland Cement Concrete," *Annual Book of ASTM Standards,* Vol. 04.02, pp. 291–293 (1988).

[4] ASTM C114, "Standard Methods for Chemical Analysis of Hydraulic Cement," *Annual Book of ASTM Standards,* Vol. 04.01, pp. 135–136 (1986).

[5] ASTM C311, "Standard Test Methods for Sampling and Testing Fly Ash or Natural Pozzolans for Use as a Mineral Admixture in Portland Cement Concrete," *Annual Book of ASTM Standards,* Vol. 04.02, pp. 182–186 (1988).

[6] ASTM C188, "Standard Test Method for Density of Hydraulic Cement," *Annual Book of ASTM Standards,* Vol. 04.01, pp. 195–197 (1986).

[7] Joshi, A., "Pozzolanic Reactions in Synthetic Fly Ashes," *Doctorate Thesis,* Iowa State University, Ames, IO, pg. 52 (1978).

[8] Ellis, W. E., Jr., "Production and Utilization of Fly Ash," *Concrete Products,* pg. 37, Oct. (1986).

[9] ASTM C595, "Standard Specification for Blended Hydraulic Cements," *Annual Book of ASTM Standards*, Vol. 04.02, pp. 279–283 (1988).

[10] Jahr, J., "Possible Health Hazards from Different Types of Amorphous Silicas," *American Society for Testing and Materials*, STP 732, pg. 210 (1979).

[11] Malhotra, V. M., Carette, G. G., "Silica Fume—A Pozzolan of New Interest for Use in Some Concretes," *Concrete Construction*, May (1982).

[12] "Manual of Aggregate and Concrete Testing," *Annual Book of ASTM Standards*, Vol. 04.02, pp. 693–719 (1988).

[13] "Cement and Concrete Terminology," *SP-19, American Concrete Institute*, Detroit, MI, pg. 143 (1988).

[14] Mehta, P. K., "Sulfate Resistance of Blended Portland Cements Containing Pozzolans and Granulated Blast Furnace Slag," *Cement, Concrete and Aggregates*, Vol. 3, No. 1, pp. 35–50 (1981).

[15] Dunstan, D. R., Jr., "A Possible Method for Identifying Fly Ashes That Will Improve the Sulfate Resistance of Concretes," *Cement, Concrete and Aggregate*, Vol. 2, No. 1, pp. 20–30 (1980).

[16] Fiskaa, O. M., "Concrete in Alum Shale," *Norwegian Geotechnical Institute, Report No. 101*, (1973).

[17] "ACI Manual of Concrete Practice. 1983, Part I—Materials and General Properties of Concrete," *American Concrete Institute*, Detroit, MI (1983).

[18] Graham, D. E., "Fly Ash and Its Use in Concrete," *NRMCA Publication No. 138*, Feb. (1972).

[19] Mather, B., "The Partial Replacement of Portland Cement in Concrete," *Paper No. 135, Corps of Engineers*, Sept. (1956).

[20] ASTM C157, "Standard Test Method for Length Change of Hardened Hydraulic-Cement Mortar and Concrete," *Annual Book of ASTM Standards*, Vol. 04.02, pp. 97–101 (1988).

[21] Dodson, V. H., "The Effect of Fly Ash on the Setting Time of Concrete—Chemical or Physical," *Proceedings, Symposium N, Materials Research Society*, pp. 166–171 (1981).

[22] Pistilli, M. F., Wintersteen, R., Cechner, R., "The Uniformity and Influence of Silica Fume from a U.S. Source on the Properties of Portland Cement Concrete," *Cement, Concrete and Aggregates*, CCAGDP, Vol. 6, No. 2, pp. 120–124, Winter (1984).

[23] Huang, Cheng-yi, Feldman, R. F., "Hydration Reactions in Portland Cement-Silica Fume Pastes," *Cement and Concrete Research*, Vol. 15, pp. 585–592 (1985).

[24] Soroka, I., Stern, N., "Effect of Calcareous Fillers on Sulfate Resistance of Portland Cement," *Ceramic Bulletin*, No. 55, pp. 594–595 (1976).

APPENDIX

Both the American Society for Testing and Materials and the American Concrete Institute require that a table of conversion factors of "inch-pound" units to SI units (or vice-versa) be included at the end of a document. Those conversion factors that apply to the preceding chapters are given in the following table.

Selected Conversion Factors

TO CONVERT FROM	TO	MULTIPLY BY
degree Fahrenheit (°F)	degree Celcius (°C)	$t°C = (t°F - 32)/1.8$
foot (ft)	meter (m)	3.048×10^{-1}
foot^2 (ft^2)	square meter (m^2)	9.290×10^{-2}
foot^3 (ft^3)	cubic meter (m^3)	2.832×10^{-2}
inch (in.)	meter (m)	2.540×10^{-2}
inch^2 (in.^{2})	square meter (m^2)	6.452×10^{-4}
ounce (U.S. fluid oz)	cubic centimeter (cm^3)	2.957×10^{-1}
pound-foot/in.^{2} (psi)	pascal (Pa)	6.895×10^{-3}
yard^3 (yd^3)	cubic meter (m^3)	7.645×10^{-1}
ton (short, 2000 lb)	kilogram (kg)	9.072×10^2
centemeter^2/gram (cm^2/g)	meter^2/kilogram (m^2/kg)	1.000×10^{-1}
micron (μm)	inch (in.)	3.937×10^{-5}
pound (lb)	kilogram (kg)	4.536×10^{-1}
pound/cubic yard (lb/yd^3)	kilogram/cubic meter (k/m^3)	5.933×10^{-1}

INDEX

INDEX

Abrams' law, 26
Acid rain, 7
Addition
  defined, 23
  functional, 23
  processing, 23
Admixture, defined, 23
Admixture, chemical. *See* Chemical
  admixture
Air entraining admixtures
  defined, 129
  relative rates of reaction with calcium
    ions, 135–136
  relative solubilities of film around air
    bubble, 133–134
  chemical classes
    anionic, 130
    cationic, 130
    nonionic, 130
  composition, 131
  history, 129–130
Air entraining cement, 9
Air entrainment in fresh concrete,
  influencing factors
  alkali in cement, 138–139
  cement, 136
  chemical admixtures, 87, 137
  coarse aggregate, 136
  concrete mixer, 137
  contaminants, 137
  fine aggregate, 136
  fineness of cement, 138
  mixing time, 137
  pozzolans, 137
  slump of concrete, 137
  temperature, 137

  vibration, 137
  water, 136
Air entrainment in fresh concrete
  coalescence of air bubbles, 132–134
  economics of use, 154–157
  effect on concrete bleeding, 144–145
  effect on W/C, 147–148
  methods of measurement, 130
  stability, effect of cement alkali, 138–
    139
Air entrainment in hardened concrete
  chord length vs. diameter of air void,
    134
  effect on compressive strength, 148
  effect of admixture composition, 134–
    135
  location of air voids in concrete, 131,
    146–147
Air entraining depressant, 62
Alite phase of portland cement, 6, 17, 77–
  78, 81
Alkali-aggregate reaction, 92
  effect of pozzolans, 185–187
  factors influencing, 184–185
  types, 184
Alkali-carbonate reaction, 184
Alkali ions in cement clinker, 6
Alkali-silica reaction, 184
Alkali-silicate reaction, 184
Aluminate hydrate, 123–124
Antifreeze additives, 98
Aragonite, 4
Aspdin, J., 2

Bauxite, 4
Belite phase of portland cement, 6